BIOMES
OF THE WORLD

VOLUME 8
Temperate Grasslands

MICHAEL ALLABY

GROLIER
EDUCATIONAL

About This Set

Biomes of the World is a nine-volume set that describes all the major landscapes (biomes) that are found across the Earth. Biomes are large areas of the world where living conditions for plants and animals are broadly similar, so that the vegetation in these locations appears much the same. Each of the books in this set describes one or more of the main biomes: Volume 1: The Polar Regions (tundra, ice cap, and permanent ice); Volume 2: Deserts (desert and semidesert); Volume 3: Oceans (oceans and islands); Volume 4: Wetlands (lakes, rivers, marshes, and estuaries); Volume 5: Mountains (mountain and highland); Volume 6: Temperate Forests (boreal coniferous forest or taiga, coastal coniferous forest, broad-leaf and mixed forest, Mediterranean forest and scrub); Volume 7: Tropical Forests (rain forest and monsoon forest); **Volume 8: Temperate Grasslands** (prairie, steppe, and pampas); Volume 9: Tropical Grasslands (savanna).

The books each have three sections. The first describes the geographical location of the biome, its climate, and other physical features that make it the way it is. The second section describes the plants and animals that inhabit the biome and the ways in which they react to each other. The final section of each book deals with the threats to the biome and what is being done to reduce these. An introduction in Volume 1 includes a map showing all the biomes described in this set, and a map showing all the countries of the world.

Throughout the pages of this set there are diagrams explaining the processes described in the text, artwork depictions of animals and plants, diagrams showing ecosystems, and tables. The many color photographs bring each biome to life. At the end of each book there is a glossary explaining the meaning of technical words used, a list of other sources of reference (books and websites), followed by an index to all the volumes in the set.

Published 1999 by Grolier Educational, Danbury, CT 06816

This edition published exclusively for the school and library market

Planned and produced by Andromeda Oxford Limited, 11–13 The Vineyard, Abingdon, Oxon OX14 3PX, UK

Project Manager: *Graham Bateman*
Editors: *Jo Newson, Penelope Isaac*
Art Editor and Designer: *Steve McCurdy*
Cartography: *Richard Watts, Tim Williams*
Editorial Assistant: *Marian Dreier*
Picture Manager: *Claire Turner*
Production: *Nicolette Colborne*

Origination by Expo Holdings Sdn Bhd, Malaysia
Printed in Hong Kong

Set ISBN 0-7172-9341-6
Volume 8 ISBN 0-7172-9349-1

Biomes of the world.
  p. cm.
  Includes indexes.
  Contents: v. 1. Polar regions -- v. 2. Deserts -- v. 3. Oceans -- v. 4. Wetlands -- v. 5. Mountains -- v. 6. Temperate forests -- v. 7. Tropical forests -- v. 8. Temperate grassland -- v. 9. Tropical grassland.
  Summary: In nine volumes, explores each of the earth's major ecological regions, defining important features, animals, and environmental issues.
  ISBN 0-7172-9341-6 (hardcover : set : alk. paper). -- ISBN 0-7172-9349-1 (hardcover : vol. 8 : alk. paper)
  1. Biotic communities--juvenile literature. 2. Life zones--Juvenile literature. 3. Ecology--Juvenile literature. [1. Biotic communities.] I. Grolier Educational (Firm)
QH541.14.B57   1999
577--dc21   98-37524
                                                      CIP
                                                       AC

Contents

The Physical World of Temperate Grasslands

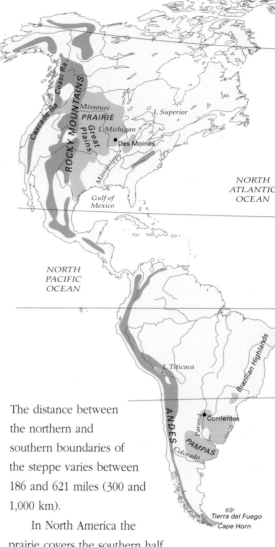

A sea of grass stretches as far as the eye can see in every direction. No mountain interrupts it, no tree breaks the surface. It is an empty place, beneath a vast, cloudless sky—and a restless place. The grass bends and sways ceaselessly in the constant wind.

In North America temperate grassland is called the prairie, in South America the pampas, in South Africa the veld, and in Europe and Asia it is known as the steppe.

Temperate grassland occupies the heart of most continents. Temperate grassland and tropical grassland (see Volume 9) together make up the vegetation that naturally covers more than one-quarter of the surface of the Earth and at one time covered more than 40 percent. Today much of it has been converted to farmland, growing mainly cereal crops, but some of the original grassland remains and is now protected.

The total area occupied by temperate grassland is immense. In Eurasia the steppe occupies the area approximately enclosed by latitudes 40°N and 50°N and longitudes 20°E and 90°E. It extends from Hungary (where it is known as *puszta*) and Ukraine all the way east to Mongolia. There are additional areas of steppe vegetation still farther east in China.

The distance between the northern and southern boundaries of the steppe varies between 186 and 621 miles (300 and 1,000 km).

In North America the prairie covers the southern half of Manitoba, Saskatchewan, and Alberta in Canada and extends southward as far as Texas. From east to west the prairie occupies the land from the eastern side of the Rocky Mountains to around the longitude of Lake Michigan. This is approximately the area between latitudes 30°N and 55°N and longitudes 100°W and 87°W.

Natural temperate grasslands are not the same everywhere. Because they extend over so immense an area, they vary considerably, and they are made up of several distinct types.

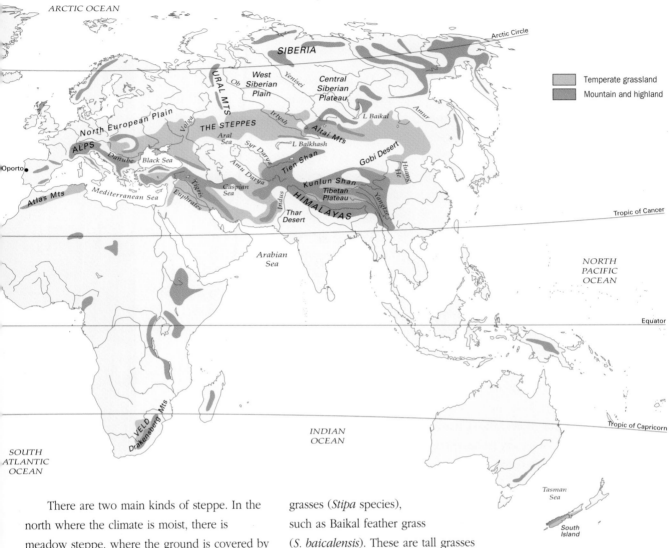

There are two main kinds of steppe. In the north where the climate is moist, there is meadow steppe, where the ground is covered by dense growths of grasses and flowering herbs. In some places 73 species of plants have been counted in one square yard (80 species in a square meter).

North of the meadow steppe the grassland begins to give way to the temperate forest known as the taiga. Trees appear on the steppe and are commoner farther north, so the grassland becomes forest steppe.

South of the meadow steppe, where the climate is drier, lies grassland known as dry steppe, where the typical plants are feather grasses (*Stipa* species), such as Baikal feather grass (*S. baicalensis*). These are tall grasses with long, feathery flowers that sway in the wind.

To its south the dry steppe gives way to desert. Vegetation becomes patchier, with an increasing proportion of desert plants. This is desert steppe.

The North American prairies are also made up of several distinct grassland types. These change as you travel across the continent. From Illinois to west of the Mississippi there is tall-grass prairie, originally with species such as big bluestem grass (*Andropogon gerardi*). West of

TEMPERATE GRASSLANDS are found where the climate is too dry for forest but too moist for desert plants. These grasslands include the prairies of North America and the veld of South Africa. The map also shows the world's principal mountain ranges.

about longitude 100°W there is short-grass prairie, with grasses such as galleta (*Hilaria jamesii*) and grama (*Bouteloua* species). Between the tall- and short-grass prairies there is mixed prairie, and further west in Washington State and British Columbia there is bunchgrass prairie. This is also called palouse prairie, after the Palouse River that runs through the region.

"Pampas" is derived from a Quechua word meaning "flat surface." (Quechua is a language spoken by some South American Indians.) South Americans usually use the singular, *la pampa*. The pampas covers an area of more than 290,000 square miles (751,000 sq. km) of Argentina, extending from the Colorado River in the south to Corrientes Des Moines in the north, and from the Atlantic Ocean in the east to the foothills of the Andes in the west.

This grassland region is made up of two distinct types. In the west the dry pampas verges on desert. The vegetation is sparse, large areas are very salty, and the rivers are brackish. In the east, where the rainfall is much higher, the pampas is fertile. This is now the center of Argentinian agriculture.

The veld—the word means "field" in Afrikaans—is on the eastern side of South Africa, bordered to the west by mountains and desert. It occupies a plateau from 4,000 feet to about 11,500 feet (1,219 to 3,505 m) above sea level. Red grass (*Themeda triandra*) occurs throughout the area, and bluegrass (*Festuca* species) grows on the higher ground.

Although the vegetation is similar everywhere, the veld is of two types, known to farmers as "sweet" and "sour." The sweet veld is on the western side. The climate is dry there, so the pasture is sparse, but the grass is nutritious at all times of year. On the sour veld to the east the climate is more moist, and the grass is very nutritious in early summer, but it is unpalatable to livestock during the winter.

SEASONS AND THE TILTED EARTH

Winter and summer are very different from each other in the temperate regions where these grasslands are found. Des Moines, Iowa, lies in the heart of the prairies, and its climate is typical.

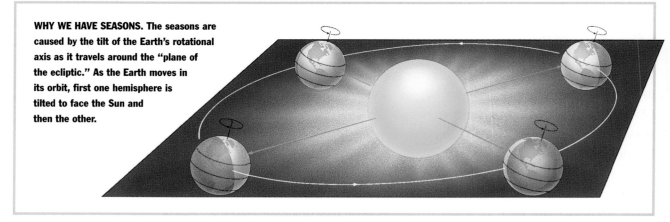

WHY WE HAVE SEASONS. The seasons are caused by the tilt of the Earth's rotational axis as it travels around the "plane of the ecliptic." As the Earth moves in its orbit, first one hemisphere is tilted to face the Sun and then the other.

Through June, July, and August the average daytime temperature is 83°F (28°C), reaching 86°F (30°C) in July. In the winter months of December, January, and February the average daytime temperature is 32°F (0°C), falling to 30°F (-1°C) in January. It rains or snows in every month, but nearly 23 inches (584 mm) of precipitation (rain, snow, sleet, and so on) fall between the beginning of April and the end of September, whereas only 9 inches (229 mm) fall from the start of October to the end of March.

Summer temperatures are higher than those in winter because the Earth is tilted on its axis. As it moves in its orbit, first one hemisphere is tilted toward the Sun and then the other.

Imagine a large, flat disk with the Sun at its center and the Earth traveling around its edge. This imaginary disk is called the plane of the ecliptic. As the Earth moves around the edge of the disk it is also spinning on its own axis. If it were spinning in an upright position, its axis would be perpendicular to the plane of the

A LITTLE HOUSE ON THE PRAIRIE in Saskatchewan, Canada. Prairie covers the southern half of the Canadian provinces of Alberta, Saskatchewan, and Manitoba, and extends southward as far as Texas.

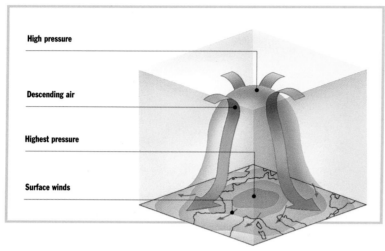

High pressure

Descending air

Highest pressure

Surface winds

ANTICYCLONES. These are regions where air is subsiding and flowing outward at the base. The subsiding air produces high atmospheric pressure at the center of the system. Anticyclones usually bring settled weather, hot in summer and cold in winter.

TORNADOES (opposite) can happen anywhere, but they are more common on the Great Plains of North America than in any other part of the world. Air is spiraling up through the funnel, generating wind speeds that in the strongest "twisters" can reach 300 mph (480 km/h). This tornado is in Oklahoma.

ecliptic, and at noon every day the Sun would be directly overhead at the equator.

The axis of the Earth is tilted, however, at an angle of 23°30'. Because of this tilt the noonday Sun is directly overhead at the equator on only two days each year. At noon on one day each year it is directly overhead at latitude 23°30'N, and at noon on one other day it is directly overhead at latitude 23°30'S. These latitudes are called the Tropics of Cancer and Capricorn respectively; the region between them is known as the tropics.

Solstices and Equinoxes

The higher the Sun rises in the sky, the more light and warmth it radiates to the surface and the longer it will remain above the horizon. It reaches its highest point at noon on Midsummer Day, when it is directly overhead at one or the other tropic. It is overhead at the Tropic of Cancer (Northern Hemisphere) on about June 21 and at the Tropic of Capricorn on about December 22. These dates mark the middle of

summer and the middle of winter in the Northern Hemisphere. In the Southern Hemisphere December 22 is Midsummer Day, and June 21 is in midwinter. These dates are known as the summer and winter solstices.

Midway between these dates, on about March 20 and September 22, the Sun is directly overhead at the equator. These dates are called respectively the spring (or vernal) and the autumnal equinoxes. "Equinox" comes from the Latin words for "equal" and "night," because at the equinoxes everywhere in the world the Sun is above the horizon for 12 hours and below it for 12 hours, so day and night are of equal length.

The time during which the Sun stays above or below the horizon reaches a maximum at the solstices, but how long it remains above or below the horizon varies with latitude. The higher the latitude—the farther away from the tropic—the longer the days are in summer, and the shorter they are in winter. The far north is known as the land of the midnight Sun because for a time in summer it is never dark. In the Arctic and Antarctic there is at least one day each year when the Sun does not sink below the horizon—it appears to travel around it—and one day when it does not rise above the horizon, so the lands of the midnight Sun can also be lands of darkness at noon!

CONTINENTAL AND MARITIME AIR

Oporto, Portugal (at 41°8'N), is in almost the same latitude as Des Moines (41°35'N), although its elevation is lower. Oporto is 312 feet (95 m) above sea level, while Des Moines is at 800 feet

(244 m). Oporto enjoys warm summers: average temperatures for June through September are 75°F (24°C), reaching 77°F (25°C) in August. In the winter months, from December through February, the average temperature is 57°F (14°C). More rain falls in winter than in summer.

The difference between the climates of Des Moines and Oporto lies in the temperature range each city experiences. In Des Moines the difference between the coldest nighttime temperature in winter and the hottest daytime temperature in summer is 74°F (41°C). In Oporto the difference is 37°F (20.5°C).

Both cities receive their weather from air that usually travels from west to east. Air reaching Oporto has crossed the Atlantic Ocean, whereas air reaching Des Moines has crossed the North American continent. Water warms and cools much more slowly than land. In summer, the land heats rapidly, but the ocean remains cool. In winter the land cools, but the ocean loses its warmth much more slowly, so by the end of the winter the sea is often warmer than land in the same latitude.

Air over a large land mass—a continent—becomes hot in summer but very cold in winter; air over a continent is also dry because there is little opportunity for water to evaporate into it. It is called a continental air mass, and its temperature, pressure, and humidity are fairly constant throughout it.

Air that crosses the ocean forms a "maritime air mass." Its temperature is related to that of the ocean; it is cooler than continental air in summer and warmer than continental air in winter. It is also moist because of its contact with the sea. Oporto, on the coast, receives about 45 inches

(1,143 mm) of rain a year, compared with 32 inches (813 mm) at Des Moines.

Twisters and Blizzards

All temperate grasslands have climates associated with continental air masses. They are fairly dry, with hot summers and cold winters. In winter cold, dense air produces high surface atmospheric pressure over North America and central Asia. This brings long spells of cold, clear, settled weather. In summer the pressure falls as the air is heated by contact with the ground, but the air is dry, and droughts are common.

The North American prairies suffer extreme weather more often than most grassland regions. Violent storms bring blizzards in winter (a blizzard is a snowstorm with winds of at least 35 mph [56 km/h] and a temperature no higher than 20°F [-7°C]). There are also more tornadoes here than anywhere else in the world.

The storms are a consequence of North American geography. In spring and summer cold, dry air that has crossed the Rockies is often moving southeast through North Dakota. At the same time, warm, dry air is moving north from Mexico. The warm, dry air meets the cold air and, because it is less dense, rises over it.

Then, warm, moist air moving north from the Gulf of Mexico meets the cold, dry air and the warm, dry air. The warm, moist air rises against the cold air but becomes trapped for a time beneath the warm, dry air. This air is cooling; when its temperature falls below that of the moist air beneath, the moist air rises rapidly. Its water vapor condenses to form huge storm clouds that extend from almost ground level to a height of 50,000 feet (15,240 m) or more. These

THE FORMATION OF SOILS *(right).* **As plants begin to grow and animals feed on them, organic matter is mixed with the mineral particles near the soil's surface. Plant roots and animals tunnel through the soil, mixing it and allowing air and water to circulate. In time, distinct layers or "horizons" develop. Grassland soils have a thick surface horizon. Some are chestnut brown in color** *(below),* **others black** *(bottom),* **because they contain a large amount of organic matter.**

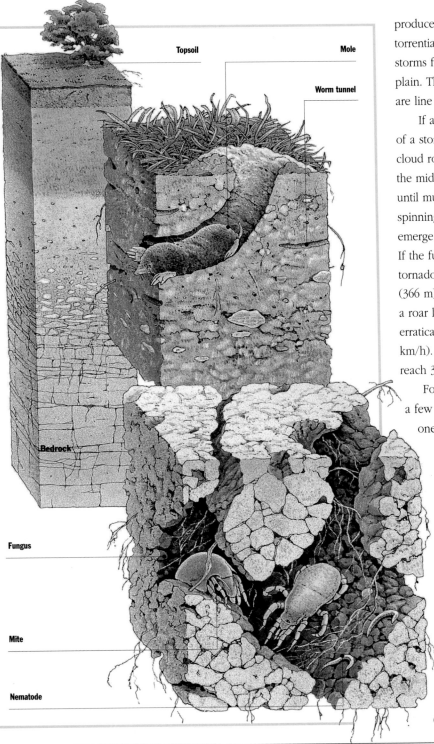

Topsoil

Mole

Worm tunnel

Bedrock

Fungus

Mite

Nematode

produce extremely violent storms with hail, torrential rain, thunder, and lightning. Often the storms form in a long line that moves across the plain. This is called a squall line, and the storms are line squalls.

If a strong wind is blowing above the tops of a storm cloud, it can set the air inside the cloud rotating. The rotation becomes strongest in the middle of the cloud, then extends downward until much of the interior of the cloud is spinning very fast. This spinning air may then emerge as a funnel below the base of the cloud. If the funnel reaches the ground, it becomes a tornado, a column of air up to about 400 yards (366 m) across at the base, swirling rapidly with a roar like an express train, and moving erratically over the ground at 25–40 mph (40–64 km/h). Inside the tornado the wind speed can reach 300 mph (483 km/h).

Fortunately, tornadoes rarely last more than a few minutes—although in 1977 there was one in the United States that lasted seven hours, during which it traveled about 340 miles (547 km) across the states of Illinois and Indiana.

ROCKS AND SOILS

Grassland soils are fertile. Cereals—which are grasses—can usually be grown in them. Beneath a dark surface layer at least 6 inches (15 cm) deep the soils are sometimes a chestnut or brown color and sometimes black. These used to be called chestnut brown soils and chernozems (the Russian for "black soil") respectively.

Nowadays there is an international system for classifying soils, and both types are placed in the soil order Mollisols.

Soil consists of tiny particles of rock mixed with plant and animal wastes in varying stages of decomposition. Its formation begins when rock near the surface of the Earth is heated by the Sun and expands, then cools and contracts. This cracks the rock. Water seeps into the cracks, expands as it freezes in winter, widening the cracks, then melts in spring and flows away. Flakes and particles of rock fall away, are rolled against one another, and are slowly worn down into fragments the size of sand grains or smaller. At the same time, rainwater washing down from above and water below ground dissolve some chemical compounds from the rocks. Further chemical reactions take place in the resulting solution, and the chemical properties of the rock particles are altered. Together these physical and chemical processes are called weathering, even though some of them are not directly linked to the weather.

Plants can use some of the ingredients of the soil solution and can anchor themselves among the particles. Their wastes, coupled with those of animals that arrive to feed on them, gradually decompose, releasing acids that cause still more reactions.

The rate at which these processes occur depends on the climate. Soil develops very slowly in climates that are too cold or too dry for plants to thrive. In temperate climates soils develop well. As they develop, distinct layers, called horizons, form in them. At the surface there is a horizon with much organic matter. Chemical compounds dissolve in rainwater and

are washed from this horizon. The process is called eluviation if it involves particles and leaching if dissolved compounds are being moved. These substances are deposited in a lower horizon. This deposition process is called illuviation. Beneath this lower, illuvial horizon there is the underlying bedrock, the surface of which is still undergoing weathering.

These are the three basic horizons, which are often called topsoil, subsoil, and bedrock. Soil scientists—the science of soils is called pedology, and soil scientists are known as pedologists—use letters of the alphabet to identify the different horizons. At the surface there is often an L horizon, of leaves, twigs, and other plant and animal material. Below it is the O horizon, where organic (carbon-containing) matter is decomposing, and below that the A horizon (the topsoil and eluvial zone) is made up of soil rich in decomposed organic matter.

The illuvial zone forms the B horizon, in which compounds are accumulating as they drain down from the topsoil. At the base of the B horizon there is rock that is undergoing weathering. This makes up the subsoil, or C, horizon. Below that the bedrock is known as the R horizon.

The O, A, and B horizons can be subdivided on the basis of their composition, using numbers: O_1, O_2, B_1, B_2, etc. B_2, for example, is where most of the illuviated compounds accumulate.

THE RECYCLING OF NUTRIENTS in the soil is a constant process that takes place through the decomposition of organic matter. The nutrients are absorbed into plant roots and used by the plants to build new stems, leaves,

Decomposers

Carnivores

Energy flow

Heat loss

make ammonia or nitrate, substances that are soluble in water and can enter plant roots. Lightning supplies the energy to combine some nitrogen with oxygen, but most nitrogen is fixed by soil bacteria. In addition to these elements organisms need phosphorus, potassium, and several more, all of which are derived from the weathering of rocks.

The dissolved nutrients are absorbed by plant roots; once inside the plant, the solution is used to construct plant tissues. The nutrients, consisting of simple chemical compounds made up of a few atoms, are assembled into much more complex organic compounds, with molecules made from many atoms. Animals cannot use the soil solution directly. They must obtain their nutrients either by eating plants or by eating plant-eating animals.

Plant and animal wastes and remains fall to the ground. They provide food for scavengers—organisms such as worms and snails. Wastes and remains from the scavengers are consumed by fungi and bacteria—the decomposers. They take the substances they need by breaking big protein and carbohydrate molecules into much smaller molecules. Then they release simple molecules that dissolve in the water present in the soil. In that form they can be taken up again by plant roots.

Some nutrients are lost from the soil, draining away in solution and entering rivers or being carried away by animals. Those losses are made good by nutrients newly weathered from the bedrock or by nutrients entering from adjacent soil. Most nutrients find their way into the sea. They may remain there a long time, but

roots, flowers, fruits, and seeds. Animals feed on these plant materials and on each other. Energy to drive these actions comes from sunlight, which green plants capture in the process called "photosynthesis."

Nutrient Cycling

Plants obtain carbon from the air, hydrogen and oxygen from water, and other chemical elements, including calcium and sulfur, from the soil. Nitrogen is needed in fairly large amounts. Air is about 78 percent nitrogen by volume, but plants cannot absorb the gas. First it must be "fixed" by making it combine with hydrogen or oxygen to

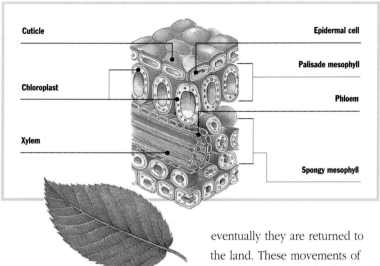

Cuticle

Epidermal cell

Palisade mesophyll

Chloroplast

Phloem

Xylem

Spongy mesophyll

PHOTOSYNTHESIS takes place in chloroplasts (above), small structures in the cells of the leaves (and sometimes stems) of green plants. A leaf has an outer, waterproof coating, the "cuticle," beneath which there is a layer of "skin" (epidermal) cells. The cells containing chloroplasts are below the epidermal cells, and beneath them are the xylem and phloem vessels conveying water and nutrient compounds from the soil, and sugars from the leaf, to other parts of the plant.

eventually they are returned to the land. These movements of nutrients on a global scale are called biogeochemical cycles.

SOLAR POWER

Within temperate grassland all the animals, fungi, and most of the microorganisms derive the energy they need to live from the food they eat.

Plants do not eat. Instead, they use water and carbon dioxide to manufacture sugars. Breaking down the sugars by combining them with oxygen releases energy. The plants use this energy to make chemical ingredients up into other substances. These substances provide nutrients to plant-eating animals (herbivores), while herbivores provide food for meat-eaters (carnivores). Plants produce food and are called producers. Animals are called consumers.

The process by which green plants manufacture sugars is called photosynthesis, and sunlight supplies the energy driving it. Plants are green because the cells in their leaves, and sometimes their stems, contain minute structures known as chloroplasts, and chloroplasts contain molecules of a chemical compound called chlorophyll. Chlorophyll is green.

When sunlight strikes a chlorophyll molecule, energy from the sunlight is absorbed, and an electron in the chlorophyll molecule passes to another chemical compound in the chloroplast. This starts a series of reactions in

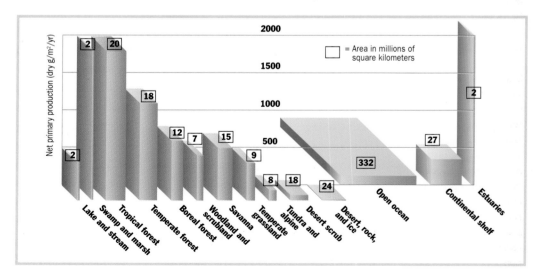

Net primary production (dry g/m²/yr)

= Area in millions of square kilometers

2000

1500

1000

500

Lake and stream — 2
Swamp and marsh — 20
Tropical forest — 18
Temperate forest — 12
Boreal forest — 7
Woodland and scrubland — 15
Savanna — 9
Temperate grassland — 8
Tundra and alpine — 18
Desert scrub — 24
Desert, rock, and ice — 332
Open ocean — 27
Continental shelf — 2
Estuaries — 2

STEPPE GRASSLANDS *(right)* on the high plateaus of Asia are very sparse, yet they support herds of grazing animals, including yaks, sheep, antelopes, gazelles, and camels.

PRODUCTION BY PHOTOSYNTHESIS *(left).* Production of living matter varies according to the climate and type of vegetation. This diagram shows the total world production in grams of dry matter per square meter per year for different types of vegetation (1 gram per square meter=0.011 ounce per square foot). It also shows the total area occupied by each type of vegetation, in millions of square kilometers (1 square km = 0.39 sq. mile), making it possible to calculate the productivity of each type for a unit area. For example, the area covered by lakes and streams is 2 million square kilometers.

which water is split into hydrogen and oxygen, and the oxygen is released into the air. This set of reactions requires sunlight, so they are called the light reactions, or light-dependent reactions.

The hydrogen then enters a second and more complicated series of reactions, sometimes called the dark reactions or light-independent reactions because they do not need sunlight to drive them. These produce a sugar, glyceraldehyde 3-phosphate. The second series of reactions form a cycle, known as the Calvin cycle because the details of the process were first described by Melvin Calvin, of the University of California at Berkeley in the United States. His discovery earned him the 1961 Nobel Prize for Chemistry.

Glyceraldehyde 3-phosphate can combine with other compounds to make proteins and fats, and it can also be converted from a sugar into a starch. Starches are the principal form in which plants store food that can be used as a source of energy.

The Natural World of Temperate Grasslands

Grasses provide food for a wide variety of animals. Herds of bison and antelopes graze on the temperate grasslands, and below ground communities of smaller animals such as prairie dogs live in networks of burrows and chambers. The animals help manage the grassland. Without them much of it would become forest.

There are about 650 genera and 9,000 species of grasses. All of them belong to the family Poaceae (also known as Gramineae). The daisy family (Asteraceae or Compositae) has about 25,000 species, the orchid family (Orchidaceae) about 18,000, and the pea family (Fabaceae or Leguminosae) about 17,000. Although these families are bigger, no plant family is more important to humans than the Gramineae. Corn (maize), wheat, barley, rice, millet, bamboo, and sugar cane are all grasses. So are the grasses grown as lawns and as pasture for livestock.

Members of this vast array of species differ in many ways, but all of them share the features that define them as grasses. They are flowering plants, angiosperms, whose name derives from the Greek *angion*, meaning "container," and *sperma*, meaning "seed." Angiosperms produce true flowers, and their seeds are enclosed in an ovary. The other major group of plants—gymnosperms—have seeds that are not enclosed in an ovary but are contained in a seed coat (their name derives from the Greek *gymnos*, meaning naked). Pines, firs, and other coniferous (cone-bearing) trees are gymnosperms.

The division (Anthophyta) containing all the flowering plants is made up of two classes, the Monocotyledoneae or monocots, and Dicotyledoneae or dicots. These names refer to the number of seed leaves, or cotyledons, that their seeds produce. Monocots produce one cotyledon; dicots product two. Photosynthesis starts in the cotyledons, which are soon replaced by true leaves.

Monocots and dicots also differ in other ways. Most dicots have leaves with veins that branch to reach every part. Monocot leaves usually have parallel veins. The bundles of vessels that transport water, nutrients, and sugars through the plant are arranged in a more complex pattern in monocots than in dicots.

Roots also differ. Many dicots have a taproot that penetrates vertically into the ground. Monocots have fibrous roots that spread to the sides. Grasses are monocots, as are palm trees, orchids, and lilies. Beans, sunflowers, roses, oak trees, and cabbages are dicots.

Recognizing Grasses

Just beneath the ground surface the roots of adjacent grass plants form tangled mats of fibrous root. Laid end to end, the roots of a single wheat plant measure more than 40 miles (64 km), and the root system of a mature corn plant (*Zea mays*) extends throughout more than 100 cubic yards (76 cu. m) of soil. The upper

horizons of grassland soils are bound together by these mats of roots, and grasses are often used to build soil on land that is being eroded.

As well as their extensive root systems many grass species also grow long underground stems, called rhizomes. Some prairie grasses, such as slender wheatgrass (*Agropyron tenerum*), produce rhizomes; others, such as buffalo grass (*Buchloe dactyloides*), have "stolons"—stems that run along the ground surface.

Grass stems are usually cylindrical and have prominent joints called nodes. Both rhizomes and stolons have branches growing from their nodes. These develop as upright stems and can also produce roots, so each node is capable of developing into a complete new plant. In species in which the stem grows vertically, it often branches at ground level. This is called tillering, and it leads to the formation of clumps or tussocks of grass. Just above each node there is a region where the cells continue to divide. This causes the stem to continue lengthening until the plant attains its full height. Other plants grow from the tips of their stems.

Leaves also originate at the nodes. The leaves of many grass species are often called blades because of their shape, although some tropical grasses have fairly broad leaves. The shape arises partly from the fact that their veins are parallel (all monocot leaves tend to be long and narrow), and partly from the way they grow.

One of the distinguishing features of all grasses is that the stem is enclosed in a cylindrical sheath. The sheath strengthens the stem and also protects the more delicate growing tissue above each node. At the top of each section of sheath there is another growing region in which the cells are dividing. It is this growth, from one half of the sheath enclosing the stem, which produces the leaf. A blade of grass is long and narrow because that is the shape of the half-sheath from which it grows.

Leaves arise singly and on alternate sides of the stem. Where the sheath and leaf join there is a rim made from a membrane, or hairs. This is called the ligule and probably serves to prevent water entering between the sheath and stem, although no one really knows its function.

An "inflorescence" is a group of individual flowers, called florets. Together the florets form what resembles a single flower. Daisies and sunflowers have inflorescences, and so do grasses; but instead of a floret, the unit of a grass inflorescence is called a spikelet. It is wind-pollinated, so has no petals to attract insects. Spikelets vary in structure from one grass species to another and are useful in identifying species. They can also be arranged in various ways. Some are attached along a single axis, like those of wheat, barley, and the prairie grass *Andropogon fastigatus* shown on page 17.

Others, like those of oats, grow on branches, forming an inflorescence called a panicle.

WHY GRASS IS SO SUCCESSFUL

Grasses are often planted to stabilize sand dunes and soils that are in danger of erosion. They are useful for this purpose because their roots bind the soil. Many perennial species—species that live for more than two years—produce rhizomes or stolons. These also bind the soil; at the same time, vertical stems and more roots branch from the nodes, allowing the plant to spread and cover the ground with a dense mass of plants.

This is why grasses are good for the soil, but it is also the way grasses have adapted to the environments in which they grow. By holding soil together, the roots trap nutrients and moisture. After a spell of dry weather, when exposed soil feels very dry, the soil held in the mat of roots beneath grass is still moist. Even if the weather is so dry that the stems and leaves die, the roots and rhizomes will survive several months longer, by

Bermuda grass
Cynodon dactylon

GRASSES have several distinctive features. Rhizomes, like those on the Bermuda grass, are stems that grow horizontally just below the surface. Grass leaves grow from the sheath enclosing the stem, as in *Poa annua*. Grass roots, such as those of the perennial ryegrass, form a mat close to the surface.

Oats
Avena sativa

Tristachya decora

fruit of the oat

Poa annua

Perennial ryegrass
Lolium perenne

which time there will probably have been enough rain for the plant to revive.

Surviving Trampling

Stems and leaves grow from the nodes at intervals along the stem. This growth pattern also benefits grasses, allowing them to tolerate trampling. This is why grasses can be grown as lawns on which people can walk, ride, and play games without causing damage.

Auxins (plant-growth substances that used to be known as plant hormones, but which are not really hormones) are produced just above each node in a grass stem, in the region where the stem is growing. Auxins stimulate plant cells to grow and divide. If the stem is upright, the auxins are distributed evenly around it, causing it to continue growing vertically upward. If the stem is flattened, however, more auxins will be produced on the underside than on the upper side. This will make cells on the underside of the flattened stem grow faster than the cells on the upper side, and the stem will grow back into an upright position.

This is helpful to the grass plant in more ways than one: the process gives grass plants an advantage over competing plants, many of which are killed by being trampled. Most plants grow from the tips of their stems and branches; if these are broken, the plant will die. Tree seedlings and young shrubs are especially vulnerable. These are plants that will otherwise grow large enough to shade the grass, and grasses are very intolerant of shade.

Surviving Grazing

In natural temperate grasslands the trampling that destroys young trees and shrubs but leaves grass unharmed is not by people but by herds of grazing animals. These animals eat grass; this is something else that grass tolerates much better than other plants.

Grass leaves (blades) grow from the upper side of nodes. The region of active growth is at the base of the blade, beside the node. If the upper part of the blade is removed, for example, by being nibbled or cut with a mower, no harm is done. The blade continues to grow. Grazing animals also eat the stems, but the stems also grow from the nodes. Growth continues from the undamaged nodes, restoring the plant to its full size. Since the lowest node is at ground level, grass will recover from even the closest nibbling.

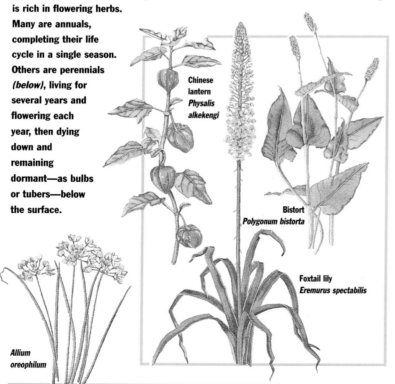

Chinese lantern
Physalis alkekengi

Bistort
Polygonum bistorta

Foxtail lily
Eremurus spectabilis

Allium oreophilum

In fact, nibbling tends to stimulate the grass to grow because it clears away old, dead leaves that shade the young, growing leaves.

Indeed, perennial grasses will recover even if the lowest node is destroyed because the upright stems and leaves are branches from the main stem that grows horizontally as a stolon or rhizome. The plant can easily survive the removal of one set of stems and leaves, and a bare patch of ground will soon be filled from the node of a nearby stem.

Other plants can survive the loss of some of their leaves, or even most of them, but they depend on them for photosynthesis and cannot continue growing if they lose too many. Nor can they survive the loss of their growing tips.

Deciduous trees and shrubs—those that shed their leaves each fall and grow new ones each spring—are especially sensitive to losing leaves. They produce leaf buds late in the year and open them in spring. These processes are stimulated by the changing length of the day, so leaves lost at other times, such as summer, cannot be replaced until the next spring.

MAINTAINING THE GRASSLANDS

Grasses dominate the vegetation in regions where the climate is too dry for trees to thrive. Provided they are not converted to farmland, these grasslands will remain indefinitely as they are today. They include the dry steppe bordering the deserts of Central Asia, the short-grass prairie in western North America, the western region of the South American pampas, and the sweet veld of South Africa.

Elsewhere, trees could grow were they given the chance. They are prevented from doing so by grazing animals aided by humans. Many centuries ago in North America Native Americans used fire to hunt game. They would set fires that would be driven by the wind, forcing herds of animals into enclosed places where they could be killed easily. Grasses were quick to appear after the fire. The surviving animals would feed on the grasses, in the process nibbling or trampling many of the young tree seedlings. Then after a few years the hunters would burn the prairie again.

This procedure was repeated at intervals over such a long period that little by little the soil lost its store of viable tree seeds. After each fire there were fewer tree seeds left to germinate (sprout), trees were never allowed to mature and produce fresh seed, and eventually the hunters and the grazing mammals succeeded in creating the prairie. Native Americans were not unique—grasslands were produced in the same way in other parts of the world.

Should the world climate become warmer in future centuries, it will also become wetter. If temperatures began to rise, more water would evaporate, more clouds would form, and precipitation would increase.

This could alter the dry grasslands, the ones that are now stable. If the ground becomes wetter, conditions could be more favorable for trees. Unless steps are taken to maintain them in their present state, shrubs could shade and slowly replace the grasses. The grasslands would become scrub, and eventually they might develop into forest grown from seed spread from the forest bordering the grassland.

This is not necessarily what would happen, though. Higher temperatures would increase the rate at which water evaporates from the soil as well as from the surface of lakes and oceans. If the amount of water that evaporated from the ground in a year were greater than the amount of precipitation, the soil would become drier, and the higher temperature would change grassland into desert—and some forests into grassland.

The Grassland Ecosystem
Grasses, herbs, and in some places scattered shrubs are the green plants of the grasslands.

PRAIRIE WETLANDS are called "sloughs." They occur in the tall-grass prairie, on poorly drained ground that is often waterlogged in spring after the snow has melted. Most of the original prairie has been plowed and the land drained, so few of the sloughs remain. Those that do survive provide valuable habitat for aquatic birds.

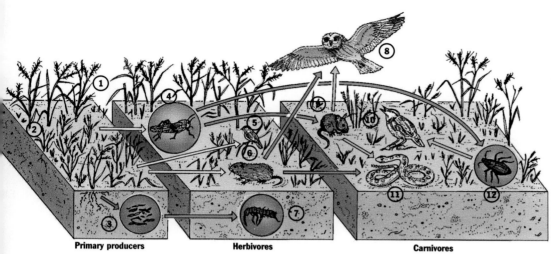

Energy flow

⟹ Primary producer/
primary consumer

➡ Primary/secondary
consumer

⟹ Secondary/tertiary
consumer

⟹ Dead material/
consumer

➡ Death

Primary producers Herbivores Carnivores

Components of the ecosystem

1 Big bluestem
2 Little bluestem
3 Detritus
4 Grasshopper
5 Grasshopper sparrow
6 Vole
7 Springtail
8 Short-eared owl
9 Deer mouse
10 Eastern meadowlark
11 Rattlesnake
12 Carabid beetle

THE PRAIRIE ECOSYSTEM. Prairie soil is rich and fertile. The grasses and herbs support a diverse population of herbivorous animals, including many insects and small mammals as well as the bison *(Bison bison)* and other large herbivores.

These are the producers on which all the animals —the consumers—depend. The food they offer consists of leaves and grass blades, roots, and seeds, and all of it is to be found at or very close to ground level.

Although the grass and other plant species vary from place to place, the type of vegetation is fairly uniform throughout the prairies, pampas, veld, and steppe. Consequently, the habitat changes little over large areas. Animal species that are able to thrive in the grassland will be able to do so in almost any part of it, and so, although there are far fewer species than in forests, the grassland ecosystem contains animal species with a wide distribution.

FLOWERS OF THE GRASSLAND

Natural grasslands are ablaze with color throughout most of the year. Nowadays, pasture is managed to make it as nutritious as possible for the farm animals that graze it. This has meant the disappearance of many wild flowers. Where the original vegetation has been left undisturbed, however, it contains many species.

Spring on the tall-grass prairie, for example, is when the lupins (*Lupinus* species) flower, as well as downy phlox (*Phlox pilosa*), prairie buttercup (*Ranunculus fascicularis*), bird's-foot violet (*Viola pedata*), prairie smoke (*Geum triflorum*), and many more. As spring turns into summer, these flowers die away, and the summer flowers take their place. In summer wild bergamot (*Monarda fistulosa*), wild indigo (*Baptisia leucantha*), leadplant (*Amorpha canescans*), and prairie coneflower (*Ratibida pinnata*) come into bloom. Summer also sees the appearance of the compass plant (*Silphium laciniatum*), which avoids exposing its leaves to the full glare of the midday Sun by holding them upright with their edges aligned approximately north–south. This exposes them to the less intense sunlight of morning and evening.

Rattlesnake master (*Eryngium yuccifolium*) is one of the plants that flowers in late summer. It earns its name because traditionally a medicine made from it was used to treat snakebites, including those of rattlesnakes. Some milkweeds (*Asclepias* species) also have medicinal uses, but whorled milkweed (*A. verticillata*), a common prairie plant, is very poisonous.

Large herbivores		Medium herbivores	Small herbivores
LOW NUTRIENTS, LOW ABUNDANCE	**LOW NUTRIENTS, HIGH ABUNDANCE**	**MEDIUM NUTRIENTS AND ABUNDANCE**	**HIGH NUTRIENTS, LOW ABUNDANCE**

The flowers continue into the fall with silky aster (*Aster sericeus*), smooth aster (*A. laevis*), rough blazing star (*Liatris aspera*), Riddell's goldenrod (*Solidago riddellii*), and many more. It is not until the first snows cover the prairie that the flowers vanish.

Flowers of the Steppe

In Central Asia, where the spring is cold and dry, there are few spring flowers. There the plants are mainly perennial, and the most typical are wormwoods (*Artemisia* species)—the plants known in North America as sagebrushes. The meadow steppe's moister climate makes it much more colorful than the dry steppe. Wild onions (*Allium* species) are very common and very diverse. *A. rotundum*, for example, has dark-colored, globe-shaped flowers that stand prominently above the grass, and *A. oreophilum* has delicate pink flowers.

Like the flowers of the prairie, the flowers of the meadow steppe appear in a succession through spring and summer. There are crocuses (*Crocus* species), tulips (*Tulipa* species), irises (*Iris* species), valerians (*Valeriana* species), and hyacinths (*Hyacinthus* species).

All size groups

**HIGH NUTRIENTS
AND ABUNDANCE**

SHARING THE RESOURCE

Grasslands support few animal species, but they do support a variety of herbivorous (plant-eating) species. In the prairie these range from grasshoppers weighing one-hundredth of an ounce (0.3 g) to bison, with females weighing an average 1,200 pounds (544 kg) and males 1,800 pounds (816 kg). If two species living side by side depend on precisely the same food, one of them will be more efficient than the other at exploiting the food resource. It may be better at

locating the food, or at gathering it, or its digestive system may work better so that its body derives more nutritional benefit from the food it eats. Members of that species will be healthier and produce more offspring than members of the less efficient rival species. After several generations the less efficient species will disappear from the habitat.

This is called the competitive exclusion principle, or Gause's principle, because it was first demonstrated experimentally by the Russian ecologist Georgyi Frantsevich Gause (1910–1986), who described it in a book called *The Struggle for Existence*, published in 1934. The principle states that two or more species cannot exist together in the same environment if all those species depend on the same resource and use it in the same way, because all but one of the species will be eliminated.

According to this principle, many animal species can feed side by side on grassland plants only if they eat different plants or different parts of the same plant. They must not compete among themselves directly. What allows them to avoid competing (and therefore to survive) is the differences in their sizes and the dietary requirements that those differences impose.

Some plants are more abundant than others, some are easier to eat than others, and some are more nutritious than others. The animals also vary. Endotherms ("warm-blooded" animals) need more food than poikilotherms ("cold-blooded" animals) because they use a great deal of energy maintaining a fairly constant body temperature. Small animals have bodies with a larger surface area in relation to their volume than big animals. They lose heat more quickly

**HOW ANIMALS COEXIST.
Animals share the food provided by the grasslands. Although bison** *(Bison bison)* *(opposite)* **are the most famous inhabitants of the prairies (this herd is in North Dakota), they are not the only animals that feed on the plants that grow there. Although some of the very plentiful plants are eaten by all the herbivores, most are sought only by certain species. This specialization means that the herbivorous species do not compete with each other.**

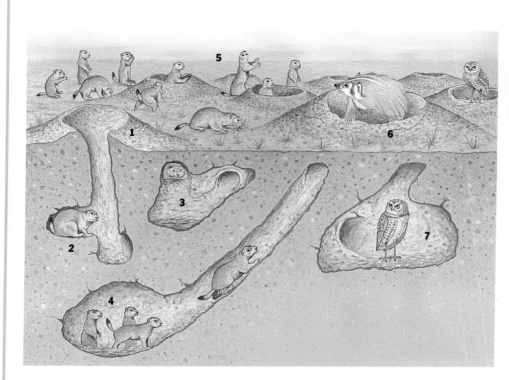

PRAIRIE DOG TOWN.
Prairie dogs are social animals that live in family groups called coteries. Several coteries—up to 1,000 animals—make up a prairie dog town.

1 Entrance dike protects against floods
2 Guard room
3 Grass-lined nest
4 New litter at end of tunnel
5 Lookouts giving warning bark
6 Badger digging for prey
7 Vacant site taken over by burrowing owls

through the larger surface area, and so they need a more nutritious diet to sustain them.

All herbivores eat plant material, but they do not have a free choice of what they eat. Insects can eat small plants or parts of plants. These do not need to be very plentiful because insects are highly mobile, and insect populations increase and decrease rapidly according to the availability of resources. Their food has to be easy to eat and very nutritious, however. Big herbivores, which do not need highly nutritious food but do need plenty of it, choose plants that are very abundant. They might also have difficulty locating tiny plants, so will eat bigger ones. The grassland herbivores can be arranged in groups of those requiring food that is nutritious but not very plentiful, plentiful but not very nutritious, and the medium-sized animals that need a diet of moderately abundant, moderately nutritious food. Some plants are both nutritious and plentiful, and all the herbivores will eat those.

RODENTS OF THE GRASSLANDS

Some herbivores actually manage their environment in ways that make their preferred foods more plentiful. Black-tailed prairie dogs (*Cynomys ludovicianus*), for example, clear away the tall plants from around the mounds from

(*C. leucurus*) is smaller and lives on higher ground in smaller colonies.

Prairie dogs are squirrels that live on the ground. Adults are about 12 inches (30 cm) long, with a tail (not a bushy one like a tree squirrel's) about 3.5 inches (9 cm) long. They are called dogs because they look a little like small terriers, and they warn each other of danger with a sharp bark, like that of a dog.

Towns below Ground

There is no shelter on the prairie, and, like many rodents, the prairie dogs live below ground. Many ground squirrels are social, but the black-tailed prairie dog is the most social of them all. Prairie dogs live as family groups called coteries; several coteries make up a "town," which might have up to 1,000 inhabitants and cover up to 160 acres (65 ha). The town is made up of a network of tunnels and chambers. The tunnels are linked, so animals can travel from one side of the town to the other without surfacing.

Mounds of earth surround the many entrances to the town. These prevent water from flooding the tunnels at times of heavy rain and also provide elevated positions from which individuals can keep a constant watch for predators, allowing the other prairie dogs to feed in peace. Just below each entrance there is a small chamber occupied by a guard that checks the returning animals to make sure there are no intruders from other coteries.

Each coterie occupies its own tunnels and chambers and has its own entrances. Some of the chambers are used for sleeping, and some, lined with grass, are where the young are born in spring.

YOUNG PRAIRIE DOGS are born naked and with their eyes closed. Their parents care for them until they are about 15 months old. These youngsters are old enough to be allowed out to search for food.

which they keep a constant lookout for predators. Tall plants would provide cover, but clearing them and keeping the ground open also encourages the growth of the smaller plants on which the prairie dogs feed. These are the prairie dogs that occur throughout the Great Plains of North America. The white-tailed prairie dog

At first the young are entirely helpless. Their eyes are closed, they have no teeth, and their skin is bare. Until they are old enough to fend for themselves, both parents collaborate in caring for them. The youngsters grow fast. After about three weeks they are fully furred; soon after the end of their first month, their eyes are open, and they have teeth. When they are about seven weeks old, their parents will allow them above ground under close supervision. Despite the lookouts, the surface is a dangerous place for young prairie dogs.

During the fall and winter members of each coterie stay within their own set of tunnels and chambers, but in spring and summer they can move between coteries. This allows breeding between members of different families.

Eventually, prairie dog towns become overcrowded. When that happens, the older animals move away and construct a new set of tunnels at the edge of town. This expands the town, but it also leaves the old burrows to the young prairie dogs, so they remain in a familiar part of the town where they know their way around, while the most experienced animals go to live as pioneers in a new district.

Bobak and Souslik of the Steppe

Burrowing rodents are the animals most typical of the steppe. The bobak or steppe marmot (*Marmota bobac*) is roughly the same size as a prairie dog. Like the prairie dog, this ground squirrel also lives in an underground colony and also manages the vegetation around it. The bobak occurs in the western part of the steppe. The Siberian marmot (*M. sibirica*) lives farther east.

Bobaks eat grass as well as a variety of other steppe plants, along with any invertebrate animals (animals lacking a backbone) they find. An adult eats up to 3 pounds (1.4 kg) of food a day and obtains all the water it needs from its food—bobaks rarely drink. In captivity they eat meat. In summer, when they feed mainly on the young shoots of grasses, their nibbling prevents the accumulation of dry, dead leaves and in that way stimulates the plants to grow. Consequently, the site of a bobak colony is visible from a long distance as a patch of green standing out against the yellows and browns of the grassland.

Like prairie dogs, bobaks keep a careful watch for predators. If one is sighted, the lookout utters piercing cries to warn the others. Bobaks are much less social than prairie dogs. Underground each animal has its own den, and there is little interaction among individuals. In September the animals move into permanent winter burrows. Once inside they block the entrances with earth and stones, and hibernate.

Sousliks are also ground squirrels. There are several species, the commonest being the little souslik (*Citellus pygmaeus*). As its name suggests, it is also one of the smallest. Adults are no more than about 8 inches (20 cm) long. They live in underground colonies, but like bobaks, each souslik has its own den. This is a chamber about 5 to 6.5 feet (1.5 to 2 m) below the surface, at the end of a sloping entrance tunnel. Sousliks

Plains pocket gopher
Geomys bursarius

POCKET GOPHERS have thick, tubular bodies and powerful limbs with large claws on their front feet. They are well equipped for a life in which digging is very important. Their incisor teeth project through their lips, so they are exposed even when the animal has its mouth closed. This allows them to bite through roots without getting soil in their mouths.

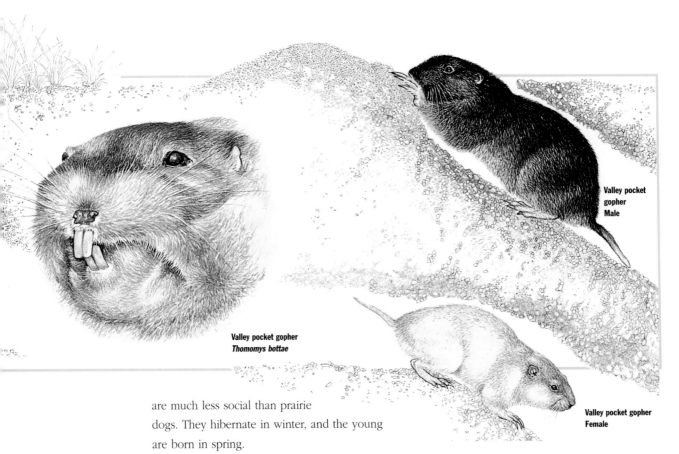

Valley pocket gopher
Male

Valley pocket gopher
Thomomys bottae

Valley pocket gopher
Female

are much less social than prairie dogs. They hibernate in winter, and the young are born in spring.

Little sousliks are one of the most serious agricultural pests in Russia. They eat crops, and their burrowing brings the salty subsoil to the surface, where it poisons pasture. They also harbor the ticks that transmit a serious disease called brucellosis to cattle.

Pocket Gophers

Wherever the prairie soil is crumbly enough to allow a small animal to make tunnels, there is likely to be a population of pocket gophers. They occur only in North and Central America, in desert as well as grassland, and they are even more highly adapted than ground squirrels to the subterranean life.

There are five genera of pocket gophers, with 34 species and many more subspecies. They range in body length from about 5 to 9 inches (13 to 23 cm) to 7 to 12 inches (18 to 30 cm), according to species. The smaller size refers to females. Males are much bigger.

The species differ in several ways, but all pocket gophers are similar in appearance. They have cheek pouches—the "pocket" of their name—in which they carry food. Their bodies are stocky and tubular and covered with very loose skin. This allows them to turn in a very small space and to move rapidly forward or backward. The tail has very little fur and is extremely sensitive to touch. Their limbs are strong, and they have very large claws on their front feet. These are their principal tools for digging. Their eyes and ears are small, but their heads are big, and they have very strong jaws.

Their incisor teeth project through their lips, so the teeth are exposed even when the animal has its mouth closed. This allows the teeth to help with the digging or to be used to cut through roots without soil entering the mouth.

They eat a wide variety of plant material, including roots and tubers they find below ground. Their tunneling mixes and aerates the soil and improves the drainage. This also tends to favor herbaceous plants at the expense of grasses—and herbaceous plants are what the gophers prefer. Like the little souslik in Russia, pocket gophers are serious agricultural pests.

Although many pocket gophers may inhabit an area with a good and reliable food supply, the animals are not in the least social. Indeed, they are fiercely territorial and will fight intruders. Each gopher occupies a territory that provides for all its needs. The size varies according to the size of the gopher and the quality of the habitat from about 50 to 285 square yards (42 to 238 sq. m).

Other Grassland Rodents

The common hamster (*Cricetus cricetus*) is the Eurasian equivalent of the pocket gopher. It is a close relative of the Syrian or golden hamster, which is the species that has been domesticated and become a popular pet, but rather bigger—10 to 12 inches (25 to 30 cm) long. It occurs throughout the steppe and is quite common, except where the natural grassland has been plowed to grow cereals.

Like pocket gophers, the hamster has cheek pouches it uses to carry food. It digs a network of tunnels, complete with escape routes, up to 8 feet (2.4 m) deep, with a nesting chamber and several storage chambers. In late summer the hamster starts filling its stores with seeds, grains, roots, and other items that will keep well. It feeds on these at intervals when it wakes briefly from its hibernation during the winter.

Dormice (family Gliridae) live throughout Europe and western Asia, and also in most of Africa south of the Sahara. They do not occur in North or South America. Like the hamster, dormice also hibernate—indeed, are renowned for their ability to sleep for up to seven months of the year. During their long sleep they are sustained by the body fat they have accumulated by eating voraciously during the summer, although they also store food for use during the late winter when they wake up intermittently.

Dormice are less strictly vegetarian than most rodents. Although they will eat nuts, berries, buds, and other vegetable items, they also eat invertebrate animals of all kinds and even eggs stolen from birds.

The muskrat or musquash (*Ondatra zibethicus*) occurs widely in North America and in Eurasia. A large relative of the voles and lemmings (subfamily Microtinae), it is up to about 14 inches (35 cm) long, with a tail about 10 inches (25 cm) long. It is semi-aquatic. Its hind feet are webbed, and it spends much of its time in water. It lives in marshes and beside rivers.

Of all the burrowing rodents the most committed to underground life are the mole rats. Zokors, or Central Asian mole rats, live in the steppes of Central Asia and China. They comprise the subfamily Myospalacinae of the Murinae, the family of Old World rats

GRASSLAND RODENTS are the commonest animals of the steppe. They include several species of dormice and other mice, as well as blind mole rats. The muskrat lives in North America.

Southern birchmouse
Sicista subtilis

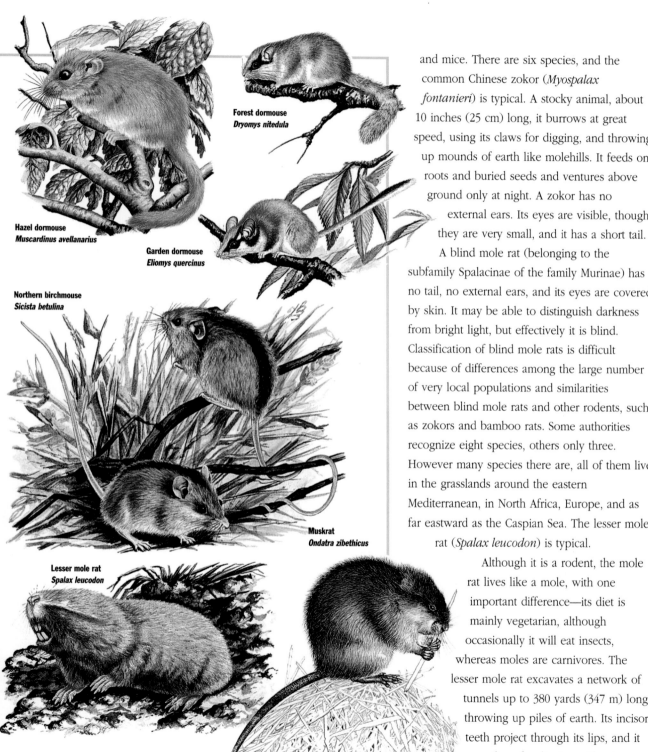

Hazel dormouse
Muscardinus avellanarius

Forest dormouse
Dryomys nitedula

Garden dormouse
Eliomys quercinus

Northern birchmouse
Sicista betulina

Muskrat
Ondatra zibethicus

Lesser mole rat
Spalax leucodon

and mice. There are six species, and the common Chinese zokor (*Myospalax fontanieri*) is typical. A stocky animal, about 10 inches (25 cm) long, it burrows at great speed, using its claws for digging, and throwing up mounds of earth like molehills. It feeds on roots and buried seeds and ventures above ground only at night. A zokor has no external ears. Its eyes are visible, though they are very small, and it has a short tail.

A blind mole rat (belonging to the subfamily Spalacinae of the family Murinae) has no tail, no external ears, and its eyes are covered by skin. It may be able to distinguish darkness from bright light, but effectively it is blind. Classification of blind mole rats is difficult because of differences among the large number of very local populations and similarities between blind mole rats and other rodents, such as zokors and bamboo rats. Some authorities recognize eight species, others only three. However many species there are, all of them live in the grasslands around the eastern Mediterranean, in North Africa, Europe, and as far eastward as the Caspian Sea. The lesser mole rat (*Spalax leucodon*) is typical.

Although it is a rodent, the mole rat lives like a mole, with one important difference—its diet is mainly vegetarian, although occasionally it will eat insects, whereas moles are carnivores. The lesser mole rat excavates a network of tunnels up to 380 yards (347 m) long, throwing up piles of earth. Its incisor teeth project through its lips, and it uses them for digging without

BUFFALO AND PRONGHORN ANTELOPE graze in herds on the North American prairies, feeding mainly on grasses and leaves.

Pronghorn antelope
Antilocapra americana

American buffalo
Bison bison

needing to open its mouth. As it moves around its tunnels, it gathers roots, bulbs, and tubers. Sometimes it seizes whole plants and drags them below ground. It stores food in chambers dug for the purpose, nests in another chamber, and even uses a special sanitary chamber. Like other burrowing animals, mole rats have benefited from the spread of agriculture; they thrive on arable land, where they have become serious pests.

PRONGHORN ANTELOPES (Antilocapra americana) grazing on the prairie in Wyoming (opposite). Several of the animals are looking toward the camera. If they sense danger, they will start running, and the whole herd will take flight, moving in long, bounding leaps at a speed that easily outruns a galloping horse.

HERDS OF GRAZING MAMMALS

Large mammals cannot burrow below ground, so they have found another way to evade predators—they live in herds. At one time buffalo herds numbered thousands of animals; the total buffalo population was about 60 million. Their meat and hides formed the basis of the economy of the Plains Indians; but as European settlers expanded westward, the

buffalo were hunted almost to extinction. William F. Cody earned the name "Buffalo Bill" because of the number of animals he killed in order to supply meat to railroad construction crews. Today there are a few thousand buffalo, most of which live in parks and wildlife refuges.

Despite its name, the American buffalo is not closely related to the buffaloes of Africa and Asia. It is a bison (*Bison bison*), related to the

European bison. There are two subspecies, the plains bison (*B. b. bison*) and wood bison (*B. b. athabascae*). Bison are wild cattle. They are big: a bull can stand about 6.5 feet (2 m) tall at the shoulder and weigh more than 1 ton (908 kg).

Animals form herds because there really is safety in numbers. Predators are sometimes confused by a large number of prey animals and hesitate while deciding which one to attack.

Among a herd of grazing animals there will always be a few whose heads are raised. These lookouts are likely to notice a predator in time to escape; when one animal is alarmed and starts to flee, the others also start running. Bison are big and strong, but although they might stand and fight, if they escape, there is no risk of injury. They swim well and can run at more than 35 mph (56 km/h).

Przewalski's horse
Equus ferus przewalskii

Pronghorn antelopes (*Antilocapra americana*) also run at that speed, and they can attain 55 mph (88 km/h) for about half a mile (800 m) as well as being able to swim. They, too, were hunted in the last century, their numbers falling from possibly as many as 50 million in 1850 to about 13,000 by 1920. They are now protected, and their numbers have increased to between 750,000 and 1 million.

They are related to the antelopes of Africa and Asia but make up a family, Antilocapridae, of which they are the only members. In summer they form small, scattered groups, but in winter they move in much bigger herds.

Saiga

The saiga (*Saiga tatarica*) lives on the steppe from Europe to Central Asia. It has excellent eyesight and can run very fast. Like the pronghorn, it was hunted almost to extinction—in this case for its horns, which are believed in China to have medicinal properties. It has been protected since 1920, and there are now more than 1 million saigas.

The saiga is a goat antelope—a close relative of the sheep and the goat, but more stockily built so that it looks more like an antelope. Only the males have horns. During summer groups of saigas feed on grass and the leaves of low shrubs. In fall they form very large herds to migrate south.

Its most remarkable feature is its nose, which is long and drooping, with nostrils that point downward. Inside each nostril there is a sac lined with mucus membranes. The nasal passages are lined with hairs and mucus glands. No one really knows why it has such an elaborate nose. It may be to filter out dust—or possibly it warms and moistens the air as it is inhaled.

Rare Species

The ancestor of all domesticated horses is an animal of the Mongolian steppe, first recognized by the Russian geographer and traveler Nikolai Mikhailovich Przewalski (1839–1888) and named after him. Przewalski's horse (*Equus ferus przewalskii*) has an upright mane, no forelock, and is reddish in color, with a pale muzzle and a dark stripe along the middle of the back and extending into the tail.

At one time these horses lived in large herds, but their numbers are greatly reduced, due mainly to competition from domesticated animals, and they now move about in small groups. Indeed, it is not certain that any survive in the wild, although small herds are maintained in some zoos.

There are still herds of wild asses on parts of the steppe. The biggest of them, standing more than 4.5 feet (1.37 m) tall at the shoulder, lives on the high grassland plateau of Tibet and Sichuan (Szechwan) in China. Its herds vary in size from five or ten animals up to as many as 400. To the Tibetans this animal is sacred. They know it as the *djang*. Outside Tibet it is called the kiang (*Equus kiang*).

The other wild ass of the temperate grasslands is called the onager (*Equus hemionus*), a name originally derived from the Greek word *onagros*, meaning "wild ass." It occurs over a large part of the Central Asian

ANIMALS OF THE GRASSLANDS include deer, wild asses, and the only surviving species of truly wild horses. Many of these species have been hunted almost to extinction—for sport, food, or for their skins. The saiga, with its distinctive long, drooping nose, was hunted for its horn, believed to have medicinal properties, but is now protected.

White-tailed deer
Odocoileus virginianus

Wapiti, elk, or red deer
Cervus elaphus

Saiga
Saiga tatarica

Asiatic ass
Equus hemionus

steppe, and there are many local subspecies, as well as local names. In Mongolia, for example, the Asian wild ass is known as the *dzhigetai*, and in Turkestan it is called the *kulan*. No one considers the onager sacred, and it has been extensively hunted for sport and for food, as well as for its skin, which is made into mats. The habitat of onagers has been greatly reduced in area as the grasslands have been turned into farms, and they have interbred with domesticated asses. There are now far fewer of them than there once were, although they survive in nature reserves. Like all asses, onagers will eat grass, shrubs such as wormwood, and thistles, but must have access to drinking water.

The Bukhara deer is also a rare species. It is a variety of wapiti (*Cervus elaphus*). At one time it roamed widely over the steppe but is now found only along some river valleys and is classed as endangered. Its American equivalent, with a range extending from southern Canada to northern South America, is not the wapiti but the white-tailed deer (*Odocoileus virginianus*). It feeds on grass, nuts, twigs, fungi, lichens—

Coyote
Canis latrans

Gray wolf
Canis lupus

COYOTES will defend food that their pack has killed and will share. In this illustration some pack members are feeding (1), while the dominant male (2) threatens a stranger attempting to intrude (3). It has adopted a defensive posture, ready to fight if it must, while another male pack member stands with a less aggressive posture behind the dominant male, giving support (4). In the distance another coyote (5) awaits the outcome of the encounter, and a group of coyotes inside their own territory (6) wait for the resident pack to finish their meal and leave, so that they can eat whatever food is left.

almost anything, in fact. It is one of the most adaptable animals in the world. It lives in small family groups, rarely forming large herds.

HUNTERS OF THE GRASSLANDS

At one time wolves (*Canis lupus*) roamed over all of North America, Europe, and Asia. Today their numbers are greatly reduced, although they are not at risk of becoming extinct. They have lost large areas of their habitat as agriculture has expanded onto more and more land, and they have been savagely persecuted. They hunt large mammals, such as bison (buffalo) and deer, although they will also eat rabbits, hares, and other smaller animals. Where farm animals graze on their hunting grounds, wolves will eat them, and very occasionally they may attack people looking after the herds or flocks. That is why

DOGS OF THE GRASSLANDS include the wolf and coyote. Like all dogs, both species are social and hunt in packs. At one time there were wolves throughout the prairie and steppes, but they have been hunted savagely, and their habitat has been reduced, so now there are few places where they are common. Coyotes occur only in North and Central America.

they have been feared and hated throughout history, and why ridding an area of wolves has always been a matter for rejoicing. In practice, however, like most carnivores, wolves prey on only the weakest members of a group—the very young or the old and sick. They rarely attack healthy adults.

On the rare occasions when they form groups to prey on large mammals, coyotes (*Canis latrans*) also hunt in this way. Most of the time, however, they catch animals no bigger than prairie dogs. Coyotes are wild dogs found only in North and Central America. They are smaller than wolves, measuring about 3 feet (91 cm) in length, not counting the tail, compared with about 4.5 feet (1.37 m) for a wolf. They are opportunists, catching whatever they can. They also eat carrion (dead flesh), as well as insects and even fruit. They have expanded their range in modern times as the decline in the number of wolves has left more prey for them.

Coyotes usually live as breeding pairs, but sometimes their young from one litter will remain with them and help care for the pups born the following year. This leads to the formation of a pack with up to eight animals. Working as a team, a pack of coyotes is able to hunt larger prey.

All dogs are social, but to varying degrees. Coyotes are less social than wolves. This may be because wolves mature in their second year and so spend longer with their parents and their litter mates than coyotes do. Coyotes mature much faster and have usually left their parents by the time they are seven months old. Coyotes often form small packs where there is abundant large prey that only a team can hunt, but live in pairs

where the most abundant food consists of small animals that cannot easily be shared. Wolves are much more committed to living in larger packs.

GRASSLAND BIRDS

One of the most beautiful of all birds is the demoiselle crane (*Anthropoides virgo*). Most cranes inhabit wetlands, but the demoiselle lives on the Eurasian steppe. Standing about 39 inches (1 m) tall, it is dark gray on the underside with white plumage on its neck, back, wings, and tail. Its numbers declined as large areas of steppe were plowed, but it has adapted to the change and now inhabits farms.

The survival of the demoiselle crane is due to its adaptability. Most cranes feed on fish, frogs, and other small animals, and also on roots they pull up from soft ground. They have big, strong bills. The demoiselle has a short bill with which it can seize insects, seeds from grasses, and the leaves of herbs. Replacing the natural grassland vegetation with farm crops amounted to only a minor change in its food supply.

The Australian grasslands are home to a relative of the cranes, the plains wanderer or collared hemipode (*Pedionomus torquatus*). Less than 7 inches (18 cm) long and with a short neck, it looks nothing like a crane. In fact it is the only member of the family Pedionomidae and related to the buttonquails (family Turnicidae). Plains wanderers run around on the ground, seldom taking to the air, although they are able to fly.

With few trees or cliffs to provide elevated nesting sites, grassland birds must nest on the

Black-tailed godwit
Limosa limosa

Lapwing
Vanellus vanellus

Meadow pipit
Anthus praetensis

Greater prairie chicken
Tympanuchus cupido

ground. Quails are plump birds, about 7 inches (18 cm) long, that eat leaves and insects. Two species occur on the steppe, the common quail (*Coturnix coturnix*), which lives throughout Europe, and the Japanese quail (*C. japonica*), found only to the east of Lake Baikal. The bobwhite quail (*Colinus virginianus*) is very similar in appearance, but slightly larger—about 10 inches (25 cm) long. It lives on the prairie. Quails spend most of their time on the ground, but they can and do fly. Quails of the northern steppes migrate south in winter.

These are game birds, hunted for food, and there are others. In Siberia there is the Daurian partridge (*Perdix dauuricae*), and on the prairie there are two species of prairie chickens. They belong to the grouse family (Tetraonidae), and like other grouse, the males perform noisy and spectacular courtship displays, inflating their orange neck sacs, raising the feathers on their necks, stamping their feet, and uttering booming cries. The greater prairie chicken (*Tympanuchus cupido*) occurs from Alberta and Montana to Texas and Louisiana, although loss of habitat has reduced its numbers. The lesser prairie chicken (*T. pallidicinctus*) inhabits the central part of the prairie, while Attwater's prairie chicken (*T. cupido attwateri*) is found on the coastal prairie of Texas and Louisiana, although it is now rare and endangered.

Megapodes are a family (Megapodiidae) that includes 12 species of birds with strong legs and very big feet—their name means "big feet." They are found only in

Horned lark
Eremophila alpestris

Australia, Southeast Asia, and islands of the Pacific. One species, the Australian brush turkey (*Alectura lathami*), lives on the Australian grasslands. Although it is hunted as a game bird (and despite its name), it is not a turkey.

Like all megapodes, the Australian brush turkey builds a remarkable nest—it is a compost heap. The male scrapes a depression in the ground, fills it with vegetation, then covers this with sand. Sunshine and the decomposition of the vegetation warm the heap, and the male tests the temperature at intervals. When it has warmed to about 86°F (30°C), the male removes the sand and makes a hole in the vegetation into which the female lays her eggs. Then he covers the heap once more, and both birds, but mainly the male, regulate the temperature by uncovering and covering the eggs as necessary. When the eggs hatch, the young birds tunnel their way out of the compost heap and immediately run away, their parents taking no further interest in them. Within a few hours the young are able to fly.

Airborne Hunters

Small animals moving through the grass are difficult to see at ground level, but the slightest movement of the vegetation is clearly visible to a bird of prey hovering or circling high above. This is the way most eagles and falcons seek their prey in open country: they really are "eagle-eyed."

Several eagles hunt over the steppe. The steppe eagle or tawny eagle (*Aquila rapax*) inhabits the western steppe, and the imperial eagle (*A. heliaca*) lives in the eastern steppe. Both are large birds, with a wingspan of about 70 inches (1.8 m).

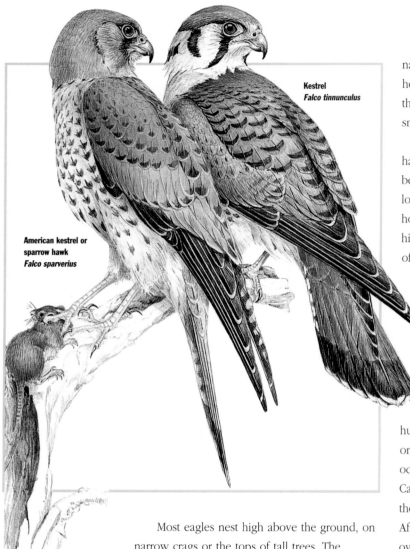

Kestrel
Falco tinnunculus

American kestrel or
sparrow hawk
Falco sparverius

**BIRDS OF PREY hunt
rodents and other small
animals by hovering while
they search for their
prey, then dropping very
fast when they locate it.
Because the grassland
habitat is very similar in
America and Eurasia, the
birds hunting there are
also very similar and
closely related.**

name suggests, it is a pale bird, with a white head and blue-gray back and wings. It nests on the ground and feeds mainly on mice and other small rodents.

Kestrels, sometimes known as sparrow hawks in North America, feed on mice and beetles, detecting them by hovering for quite long periods not far above the ground. When not hovering, they are often to be seen perched on a high lookout post, such as a tree or rock. Those of the prairie and steppe are very similar. The American kestrel (*Falco sparverius*) and kestrel (*F. tinnunculus*) differ slightly in their markings. Both are about 14 inches (36 cm) long. The lesser kestrel (*F. naumanni*) is smaller and feeds mainly on insects it catches in flight. It lives over the western steppe.

Owls also feed on small animals; most hunt only at night. The barn owl (*Tyto alba*) is one of the most widely distributed of all birds. It occurs throughout the United States (but not Canada), Central and South America except for the Amazon Basin, Europe, Arabia, most of Africa, South Asia, and Australia, but is not found over the Eurasian steppe. With so large a range there are many local variations in its appearance, but all barn owls are pale and have a big, heart-shaped facial disk. In fact, the disk resembles two parabolic dishes and is presumed to collect

Most eagles nest high above the ground, on narrow crags or the tops of tall trees. The imperial eagle builds its large nest in the top of a tree and hunts its prey by soaring. The steppe eagle is different. It nests on the ground or on a mound raised a little higher than the surrounding area, and spends much of its time standing on the ground. When it hunts, it does so at low level. Both feed on small animals such as sousliks and marmots.

Harriers are much smaller birds. The pallid harrier (*Circus macrourus*) of the western part of the steppe is about 19 inches (48 cm) long. As its

**THE BARN OWL *(Tyto alba)* hunts over open grassland
and agricultural land and occurs throughout the
prairies and pampas and in western Europe and North
Africa, but not in the steppe grasslands. Its facial disk
focuses sound to its ears (which are hidden), allowing it
to locate prey in total darkness. It is an important and
valuable predator of farm pests.**

RATTLESNAKES inhabit the prairie. This western or prairie rattlesnake (*Crotalus viridis*) hides by a low bush awaiting its prey. It feeds mainly on small mammals up to the size of hares, striking rapidly, injecting its venom, then waiting for the victim to die.

sound and focus it onto the ears, which are located beneath the disk. Its eyes are smaller than those of some owls. These features adapt the bird for hunting at night, when it locates its prey mainly by sound. In fact, a barn owl can find and capture prey in total darkness. Barn owls are very popular birds, especially with farmers, because they kill substantial numbers of rats, mice, and other agricultural pests.

REPTILES AND AMPHIBIANS

Birds and dogs are not the only predators of small grassland animals. There are also snakes that feed on small mammals, birds, and other snakes, and lizards that feed mainly on invertebrate animals. Frogs and salamanders prey mainly on worms and insects, but the tiger salamander (*Ambystoma tigrinum*) also eats mice and other vertebrate animals.

Salamanders and frogs are amphibians, members of the class Amphibia. Amphibians were the first vertebrate animals—animals with a skull and spine—to live on land, but the larvae of almost all species of amphibians are aquatic, so these animals must return to water to reproduce. As land-dwelling adults, most amphibians possess lungs, but they also absorb oxygen and release carbon dioxide through their skins. For this to work the skin must be kept moist. Consequently, most amphibians are unable to venture far from water.

Breathing through their skins—this is called cutaneous respiration—also limits the size to which an amphibian can grow: beyond a certain size the body cannot be sustained by cutaneous respiration.

The tiger salamander is the biggest land-dwelling salamander in the world, measuring up to about 16 inches (41 cm) in length. It lives on the prairies, everywhere from Canada to Mexico. It inhabits the damp vegetation near rivers or wet ground and is one of about 32 species of mole salamanders (family Ambystomatidae), all of which are found only in North America. "Mole" salamanders are so-called because of their habit of living in a burrow. They emerge at night, especially after it has rained. Like all amphibians, a tiger salamander will eat any animal it can catch. Its own size determines the size of its prey. A tiger salamander will catch and eat an animal the size of a mouse.

In the western part of their range tiger salamanders are able to breed while they are still aquatic larvae. Those living in the east of North America, however, must develop into land-dwelling adults before they are able to breed.

Rattlesnakes, Lizards, and Tortoises

The western or prairie rattlesnake (*Crotalus viridis*) inhabits the prairies. Like all rattlesnakes, it is a pit viper, possessing organs set into small depressions—pits—just below its eyes. These sense minute changes in temperature and so allow the snake to detect the presence of an animal. This means pit vipers can hunt at night and can pursue prey into the darkest of corners and burrows. The western or prairie rattlesnake feeds on animals up to the size of hares.

AMPHIBIANS AND REPTILES of the prairie include salamanders, lizards, and tortoises. Salamanders must remain moist. They live near water and forage for food at night.

Gopher tortoises
Gopherus polyphemus

Tiger salamander
Ambystoma tigrinum

Texas horned lizard
Phrynosoma cornutum

Green anole
Anolis carolinensis

The green anole (*Anolis carolinensis*) inhabits the southern United States, from Texas to Florida and Virginia. It can grow up to about 8 inches (20 cm) long and eats insects and spiders (most lizards are carnivores). Although it is ordinarily green, like many anoles it can change color, turning brown in a matter of seconds. The curious pink flap on the throat of the male is used in courtship display.

The green anole is related to the iguana (family Iguanidae), as is the Texas horned lizard (*Phrynosoma cornutum*), an animal up to 7 inches (18 cm) long with a body covered in defensive armor. The lizard feeds mainly on ants and spends much of its time hidden among vegetation or buried beneath loose soil.

As the "gopher" in its name suggests, the gopher tortoise (*Gopherus polyphemus*) is also a burrower and has front legs that are flattened like shovels. Each tortoise excavates a tunnel, sometimes more than 40 feet (12 m) long, ending in a chamber where the temperature and humidity remain fairly constant. It emerges during the day to bask in the sunshine and feed on grass and leaves. Gopher tortoises live in the southeastern United States.

WILDLIFE ON FARMS

The plants on arable farms, such as corn (maize), wheat, potatoes and beet, are not very different from those of the natural grassland. Indeed, wheat, barley, oats, and rye are descended from grasses that grow naturally on the southern steppes of Eurasia. Farms that raise cattle and sheep, feeding them on grass, also have pasture that may have been grown from seed or be made up of plants that grow naturally, perhaps with the help of fertilizer to stimulate growth. Cultivated or managed pasture resembles natural grassland even more closely than arable crops do. It is not surprising, therefore, that some birds and mammals have been able to adapt to this change in their habitat, and some have benefited. Cultivating the soil and plowing in the residues of each crop increases the population of the earthworms and other small soil animals on which many larger animals feed. Others eat seeds—supplied in abundance by the farmer.

FARMING AND WILDLIFE. As forests and other areas have been cleared and plowed, the farms that replace them are not very different from grassland. Substituting arable crops for natural grasslands is only a minor change for the animals that feed on grassland plants. Consequently, many animals have adapted and now thrive on farms.

Virginia opossum
Didelphis virginiana

Asian white-backed vulture
Gyps bengalensis

Harvest mouse
Micromys minutus

Red-legged partridge
Alectoris rufa

Paddyfield warbler
Acrocephalus agricola

Cane rat
Thryonomys gregorianus

Some of the wild species that live on farms feed mainly by scavenging. The common or Virginia opossum (*Didelphis virginiana*) feeds on insects, carrion, fruit, and whatever it can retrieve from garbage cans. About the size of a cat, but with shorter legs, it is the only native North American marsupial mammal.

Animals that remove dead carcasses quickly are performing a useful service. Vultures are the most famous of all scavengers. The Asian white-backed vulture (*Gyps bengalensis*) specializes in animals that are permanent residents rather than migrants. So the spread of farming, based on large numbers of animals that do not travel far, suits its way of life. It is one of the smallest vultures, measuring only about 3 feet (91 cm) long, and it nests in trees near villages.

Partridges (family Phasianidae) are game birds, hunted for sport and food. They are plump and about 14 inches (36 cm) long. Partridges live on the ground and prefer to run rather than fly if disturbed. They feed on plants and insects and often make their nests among farm crops.

The red-legged partridge (*Alectoris rufa*), found in Spain, Portugal, southwestern France, and southern England, where it is common on farmland, is unusual in that the female lays two clutches of eggs: one she incubates herself, the other is incubated by the male. The gray partridge (*Perdix perdix*) has a wider distribution across Eurasia, and also occurs in North America.

Warblers are made up of two subfamilies (Old World warblers are Sylviinae; Australian warblers are Acanthizinae) in the family Muscicapidae. They are small, insect-eating birds, many of which inhabit wetlands, nesting among reeds and sedges. Agriculture has not benefited most of them, but there is one notable exception. In Southeast Asia rice cultivation has greatly aided the paddyfield warbler (*Acrocephalus agricola*). As its name suggests, it is perfectly at home in the rice fields (*agricola*—like agriculture—comes from the Latin words for field and cultivate).

Shrews also eat insects and other invertebrate animals, and various species are found throughout North America and Eurasia, living wherever there is vegetation to provide cover. The Eurasian common shrew (*Sorex araneus*) is very plentiful throughout Europe and as far east as the Yenisei River. It measures no more than about 3.5 inches (9 cm) long, not including the tail. The pygmy shrew (*S. minutus*) is only 2.5 inches (6 cm) long. Because they are so small, they have a large surface area (through which heat can be lost) in relation to their volume and must eat frequently to maintain their body temperature. This explains their reputation for having a voracious appetite. Their other reputation, for bad temper, stems from their need to defend their territories, which they do vigorously. Encounters between shrews are always aggressive, except when males and females meet to mate. Shrews are so numerous that together they consume large numbers of insects. Some of these are crop pests, so shrews are popular with farmers.

Farm Pests

Moles, on the other hand, are regarded as pests. They are found throughout the prairie and steppe as well as in western Europe. The European mole (*Talpa europaea*), North American eastern mole (*Scalopus aquaticus*), and western mole (*S. latimanus*) are very similar.

They live underground, surfacing only at night, and eat worms, insect larvae, and other invertebrates that fall into their elaborate network of tunnels. In this way they destroy many pests, such as those attacking plant roots, but in doing so they also harm the roots themselves. In Russia the little souslik (*Citellus pygmaeus*) is a pest for the same reason.

Moles are found mainly in grassland, including cultivated pasture and garden lawns. They are uncommon in arable fields, because plowing destroys their tunnels.

One of the smallest of all mice, the harvest mouse (*Micromys minutus*) is only about 3 inches (7.6 cm) long, with a tail about the same length. It lives over most of Europe and Asia, and there is some local variation—those living in Russia are bigger than the ones found in Britain. It does not occur in North America.

Harvest mice make nests of grass blades among vegetation well clear of the ground, often around the edges of fields. Grassland animals, they have established themselves on farms throughout their range. They eat insects— they have been seen to chase flies and moths— as well as berries, other fruit, and leaves. They also eat cereal grains, climbing the stalk of the plant to obtain them. Harvest mice do not cause too much harm in western Europe, but they are serious pests on Russian farms.

MANY MAMMALS thrive on modern farms. Some, such as shrews, have no effect on the farm itself, but others cause considerable harm. The rabbit is one of the most serious of all farm pests, and hares also cause some damage. Moles damage plant roots by tunneling beneath them. Sousliks also tunnel and will eat crops as well.

Rabbit
Oryctolagus cuniculus

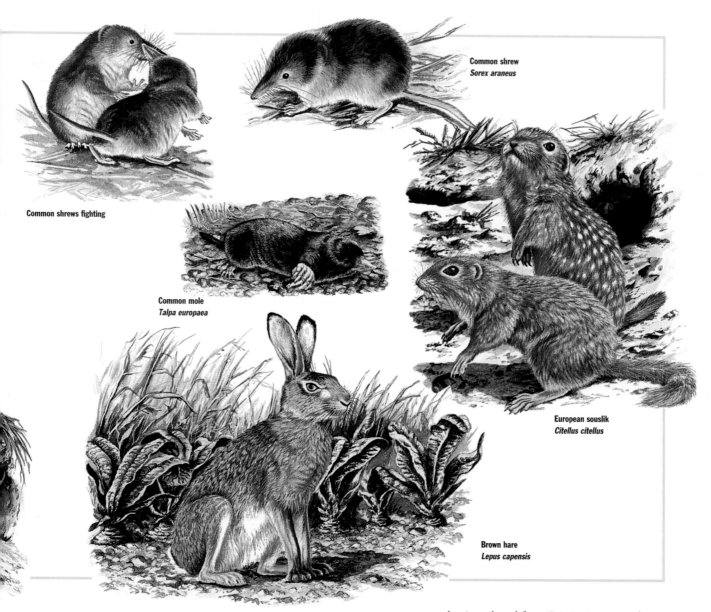

Common shrew
Sorex araneus

Common shrews fighting

Common mole
Talpa europaea

European souslik
Citellus citellus

Brown hare
Lepus capensis

The most serious of all mammal pests is the rabbit (*Oryctolagus cuniculus*). Originally a native of North Africa, during the Middle Ages it was introduced to Spain and from there to the rest of western Europe and was kept as a source of meat and fur. In the last century the rabbit was also introduced from Britain into Australia, New Zealand, North America, and Chile. In the absence of most of the predators that hunt it in Africa, and provided with ample grassland on which to feed and burrow, it proliferated in all these countries and was soon causing immense damage to crops, especially pasture.

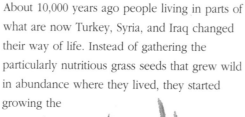

Survival of the Temperate Grasslands

With their deep, dark, fertile soils, the temperate grasslands are easily converted to farms. Only a few trees need to be cleared before the land can be plowed; livestock can graze on any land unsuitable for cropping. In North America, South America, South Africa, Europe, and Russia vast areas of the prairies, pampas, veld, and steppe have vanished. Today this land is farmed or ranched.

OUR MOST IMPORTANT STAPLE FOODS are cereals (grasses that produce edible grains). The world produces more wheat and rice than any other food; these grasses have been cultivated for many thousands of years. Barley is used mainly to feed livestock.

About 10,000 years ago people living in parts of what are now Turkey, Syria, and Iraq changed their way of life. Instead of gathering the particularly nutritious grass seeds that grew wild in abundance where they lived, they started growing the grasses that

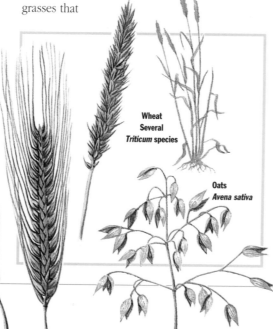

Rye
Secale cereale

Barley
Hordeum vulgare

Wheat
Several
Triticum species

Oats
Avena sativa

produced them. At other times and in other places—in Central America, India, and China, for example—other people did the same with the edible plants on which they relied for food.

Around the same time, other people began capturing animals and raising them, eventually allowing them to breed under controlled conditions rather than hunting them. The cattle, sheep, and goats that supplied them with milk, meat, and skins were fed on natural pastures, and the herds and flocks were moved with the seasons so that they could gain access to the most nutritious food.

These were the most momentous changes in all of human history. They marked the beginning of farming and livestock husbandry, and they led to vast increases in the food supply. Those increases are still continuing. Between 1994 and 1995 and the estimated harvest for 1997–1998 world production of wheat increased by 15 percent and rice production by 5 percent.

Most of our cereals are grown on land that was once grassland. Forests have been cleared to make fields, but temperate forests usually grow in climates that are too cool, too cloudy, and too wet for cereals to ripen reliably. So although there are exceptions, in general forest is cleared to make way for pasture, for cereals grown to feed livestock, and for vegetable crops.

WILD GRASSES, DOMESTICATED CEREALS

The wild grasses that were domesticated are annuals. They complete their life cycle, from the germination of seed to the setting of new seed, within one year. Once the seed has been

produced, the plant dies. This habit made cereal grasses attractive to those who began cultivating them because they produced abundant seed every year, and the seeds were nutritious.

Seed production is the means by which a plant reproduces itself. If reproduction is to succeed, the seeds must be scattered as soon as they are ripe. For this reason, ripe seeds are easily detached from the parent plant. This is all very well for the plant, but for people gathering the seeds it is highly wasteful. Grains cannot be

collected until they are ripe, and the only way to collect them is by cutting and gathering the stalk and seed head. This scatters the seeds, however, so by the time it has been carried back to the camp or village, the crop consists of a lot of empty heads and only a few remaining grains—the grains that were more firmly attached than the others.

When people started growing their own cereals, they selected these seeds that did not fall easily from the head. Over succeeding

WHEAT FIELDS cover huge areas in temperate regions. In most places they grow on land that was once natural grassland. Flour, for making bread, cakes, and pasta, is obtained by grinding wheat grains.

CEREAL FARMS ON THE FORMER EURASIAN STEPPE. The farms lie mainly in the west and stretch from the Black Sea to the Ob River. To the north they are bounded by the taiga, the northern coniferous forest. The southern steppe is uncultivated pasture land.

Agricultural zones

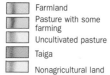

Farmland
Pasture with some farming
Uncultivated pasture
Taiga
Nonagricultural land

▲ Mountain peak (meters)
▼ Depression (meters)

generations this led to the development of cereals that do not drop their grains.

Several wild grass species were cultivated and crossbred with each other, which led to the varieties of wheat farmers grow today. Scientists are still working to trace the steps by which modern wheat was developed. What they do know is that the modern plants are genetically very different from their wild ancestors. Like all crop plants, they have been extensively modified genetically.

Some cereals were domesticated by accident. Wheat and barley were cultivated, but other seeds were mixed with theirs and grew among the crops as weeds. The weeds included rye and oats, grasses that did not grow particularly well in warm climates but flourished when cereal growing spread into northern

Europe. There they were cultivated. Even today wild oat (*Avena fatua*), a species with no value as food, survives as a troublesome weed of cereal crops.

PLOWING THE STEPPE

The deep, rich soils of the European grasslands were brought under cultivation many centuries ago, but the lands of the true steppe were plowed much more recently. Throughout its history Russia has been attacked repeatedly by nomadic warriors from Central Asia; the steppe, with no natural frontier, was not a safe place. There was no major colonization of Siberia east of the Ural Mountains and the

ARCTIC OCEAN

Wrangel Island

Chukot Range

Komsomolets
October Revolution
Bolshevik

New Siberian Islands

Koryak Range

Severnaya Zemlya

Arctic Circle

Baltic Sea

Kola Peninsula

Novaya Zemlya

Byrranga Mts.

Cherskogo Range

Pobeda ▲ 3147

Kolyma Range

RUSSIA

L. Ladoga

Yamal Peninsula

Kamchatka Peninsula
Klyuchevskaya ▲ 4750

LITHUANIA
ESTONIA
LATVIA

L. Onega

N. Dvina

Gydanskiy Peninsula

Yenisei

Verkhoyansk Range

BELORUSSIA

Central Siberian Plateau

Lena

Dzhugdzhur Range

UKRAINE

URAL MOUNTAINS

Ob

West Siberian Plain

Sea of Okhotsk

Dnieper

Sakhalin

Black Sea

Sea of Azov

Don

Volga

Ural

Tobol

Ob

RUSSIA

Stanovoy Range

Amur

Ussuri

Elbrus ▲ 5633

Caspian Depression

1
Caucasus Mts
3
2
2

Kirgiz Steppe

Irtysh

L. Baikal

Yablonovy Range

PACI
OCE

Mangyshlak Peninsula ▼ -132

KAZAKHSTAN

Eastern Sayan

Caspian Sea

Aral Sea

Western Sayan

L. Zaisan

Syr Darya

4 Kara Kum

5 Kyzyl Kum

Amu Darya

L. Balkhash

Altai

MONGOLIA

Gobi

7 Pobedy Peak
Communism Peak ▲ 7439
▲ 7495
Pamir
6

1. Georgia
2. Azerbaijan
3. Armenia
4. Turkmenistan
5. Uzbekistan
6. Tajikistan
7. Kyrgyzstan

Volga River until the 19th century, and even then most of the Russians in Siberia had been forcibly exiled there. To this day the population of Russia is concentrated mainly in the west.

Agriculture succeeded on the black soil—known as chernozem—of the western steppe, but harvest failures were frequent and followed by famines. By the middle of this century agriculture was still the weak link in the national economy. Agricultural modernization, with the introduction of machines and fertilizers, began late, and the sheer size of the country caused problems. Raw materials and harvested produce had to be transported huge distances.

An attempt began in the 1950s to increase agricultural production and at the same time to encourage people to move to the east and south of what was then the Soviet Union. Large areas of virgin steppe in western Siberia and northern Kazakhstan were plowed, the land irrigated, and cereal crops sown. It was not long before trouble began.

Prairie and steppe grasses are perennials. They do not die at the end of every season. Most species spread by means of rhizomes and stolons rather than by seeding, although they do produce flowers and seeds. Because the plants are perennials, the ground is never bare. Grass dies down to form a brown mat of dead vegetation, but the soil surface is protected, and soil is bound together by the grass roots. After harvesting an annual crop, the ground is left bare

HARVESTING ON THE STEPPE *(left),* where a single field extends almost to the horizon. Not all the steppe is as fertile as this land, and attempts to increase food production by plowing virgin steppe proved an expensive failure.

until it has been tilled, and the following crop is established. Where the climate is moderate, this is satisfactory, but the climate of the virgin steppe was not moderate.

In winter low-pressure weather systems move across the steppe from west to east, drawing bitterly cold air down from the Arctic. The cold air arrives in the form of winds blowing with hurricane force. This type of storm is called the buran. On the dry steppe it blows away the thin covering of snow, producing violent blizzards, but it also began eroding the soil from the plowed, bare soil.

There were other problems, too. The soil in this area was not the black soil of the west, but a chestnut soil. This is also fertile, but its rich topsoil lies over subsoil that has been made alkaline by mineral salts. Even with irrigation, conditions were often very dry, and the irrigation water caused salts to rise from the subsoil and injure the crop plants. One harvest in four failed. The 1930s tragedy of the Dust Bowl in the United States (see below) was being repeated. Today trees have been planted to shelter the land from the wind, and some of it has been abandoned as farmland and restored as steppe.

GRAIN-GROWING IN NORTH AMERICA *(below right)*. Grain is grown over much of what was once the North American prairie. Spring wheat is sown in spring. Farther south, where the winters are not so long or severe, winter wheat is grown. It is sown in the fall, germinates before the snows fall, then resumes its growth the following spring. Winter wheat produces bigger yields than spring wheat. Between these two wheat areas is the "Corn Belt," where corn (maize) is grown.

Major grain-growing areas

- Spring wheat
- Winter wheat
- Corn (maize)

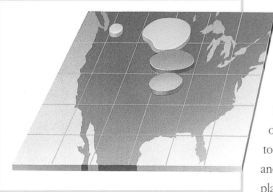

THE DUST BOWL

Americans began to farm the prairies on a large scale in the middle of the 19th century. In 1862 the Homestead Act was passed, offering free land to those who were prepared to farm it. Many people seized the opportunity and moved west. There they discovered a vast plain covered with grass that looked to them

PRAIRIE FARMING relies on a high degree of mechanization. Powerful tractors and combine harvesters are used to farm the huge fields; larger fields enable farmers to produce food more cheaply.

much like the grass that grew back home in the east. In fact the grasses were those found in a much drier climate.

The newcomers burned the natural grass and plowed the land. On the whole, crops were good. Every few years there would be a drought, and the harvest would be lost, but always the rains returned and the farms recovered. Then, from about 1910, prairie farmers began using tractors. These machines allowed them to work

much faster, so they could till a bigger area, and more of the prairie was brought into cultivation. When the Wall Street crash of September 1929 was followed by the Great Depression, cereal prices fell, but the farmers were able to work the land harder and produce more. Rainfall was above average between 1927 and 1933.

Then there came a change in the weather, and the annual rainfall decreased. Crops began to fail, and the dry soil, worked to a fine texture by the farmers, turned to dust and began to blow away in the wind. The drought began in 1933 and did not end until late in 1940. The dust storms were so severe that in May 1934 dust was settling on the desk of President Franklin D. Roosevelt, in the White House, as fast as it could be swept away. Dust settled on ships 300 miles (480 km) from the American coast. Huge dust clouds covered most of the interior of the country, and soil was blown into great dunes.

The most severely affected area came to be known as the Dust Bowl. It covered 150,000 square miles (388,500 sq. km) in Colorado, Kansas, New Mexico, northern Texas, and the Texas and Oklahoma panhandles. Farming families were ruined. They abandoned their land and homes and migrated to other parts of the country, especially to California, in the hope of rebuilding their lives.

They were victims of the first severe drought since the land had been brought into cultivation, but there have been other droughts before and since. They tend to occur at intervals of 20 to 23 years and have been recorded for nearly 170 years. Since the Dust Bowl drought there has been drought in the 1950s, 1970s, and 1990s. Scientists realized that some parts of the prairie

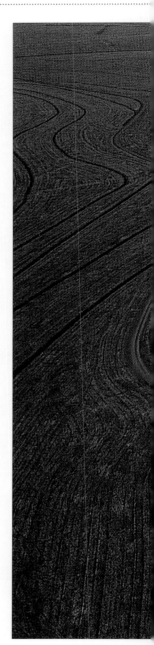

could not be cultivated safely. Prairie grasses can survive on them, because during droughts their roots bind the soil into lumps as hard as concrete, which fall apart only when the rains return to wet them. For this reason the most vulnerable areas have been encouraged to revert to prairie, and elsewhere shelter-belts of trees are grown to protect the land from the wind.

FARMING THE GRASSLANDS

Every year the Canadian prairie provinces of Manitoba, Saskatchewan, and Alberta produce more than 25 million tons (23 million tonnes) of wheat, about 15 million tons (13 million tonnes) of barley, and 7.7 million tons (7 million tonnes) of corn (maize). There, on the northern prairie, cereal farming is successful and productive. Elsewhere in the world other temperate grasslands also make major contributions to national economies.

The pampas provide a substantial proportion of the income of Argentina. The country grows more than 19 million tons (17.6 million tonnes) of wheat each year, as well as large quantities of soybeans and corn. Livestock are also raised on the pampas. Argentina has 54 million head of cattle and 17 million sheep.

The South African veld produces cattle, corn, wheat, and other crops. The country produces more than 15 million tons (13.6 million tonnes) of cereal grains a year and farms 13 million head of cattle.

New Zealand also produces arable crops, but most of its grassland is used to raise livestock. About 9 million head of cattle and 49 million sheep graze on New Zealand pastures, producing butter, cheese, meat, and wool that are the country's most important exports.

Temperate grasslands are too productive not to be farmed. It is only near their edges, where the climate is drier, that cultivation is unwise. Even there it is often possible to raise cattle on the open range. This is a type of farming that involves very little interference with the natural vegetation. The steppes, prairies, pampas, and veld support herds of grazing mammals, such as bison, pronghorns, and saiga antelopes. Ranchers merely substitute domesticated livestock and let them graze on the existing pasture.

Surviving Grasslands

Despite the attractions of farming and ranching, areas of natural grassland still remain. In the United States parts of the prairie have survived, and others have been restored for their scientific and conservation value.

In Eurasia areas of steppe that were plowed during the attempted agricultural expansion of the 1950s have been allowed to revert to their natural state. East of Lake Baikal there are large areas of the Siberian steppe that have never been plowed. Some areas have been made into nature reserves. One of the most important was established in 1874, prior to the main agricultural expansion. The Askania-Nova Reserve on the north coast of the Black Sea covers 43 square miles (111 sq. km), of which only 6 square miles (15 sq. km) has ever been plowed. The reserve is steppe that is dominated by sheep's fescue (*Festuca ovina*) and feather grass (*Stipa* species). It contains a total of 417 plant species, some of them endangered.

Crop yields have increased steadily. Today the industrialized countries of North America, western Europe, and New Zealand produce more food than consumers need. As agricultural technology continues to advance, we can expect still higher yields in the future. By producing more food from a smaller land area, for the first time in history it should be possible to reduce the area of land committed to food production. Then, little by little, what was once natural grassland may be permitted to become grassland once more. Most of the natural temperate grasslands of the world have been lost, but perhaps they have not been lost forever.

A BRAZILIAN FARM on which soybeans are being grown. This land was formerly subtropical grassland. Farming improves the soil by adding plant nutrients and organic matter.

Glossary

air mass Air covering a large area, such as a continent or ocean, that has characteristics of temperature, pressure, and humidity that are derived from the surface over which it originates, and that are fairly uniform throughout.

alga A simple green plant that lacks true leaves, stem, and root. Many algae are single-celled; some are multicelled. Seaweeds are algae.

amphibian A vertebrate animal of the class Amphibia. The young develop in water, although the adults may live on land. Frogs, toads, newts, and salamanders are amphibians.

angiosperm A flowering plant in which the ovule bearing the seeds is enclosed within an ovary.

auxin One of several plant growth substances (sometimes called plant hormones, although their action is different from that of animal hormones) that is produced at the growing tips of stems and roots.

bacteria Microscopic organisms, most of which are single-celled, that are found in air, water, and soil everywhere. Different types vary in shape and way of life.

biogeochemical cycle The movement of chemical elements from rocks and soil through living organisms, water, and air in an approximately cyclical pattern. If the elements are plant nutrients, the cycle is known as a "nutrient cycle."

biome A large region throughout which living conditions for plants and animals are broadly similar, so the region can be classified according to its vegetation type.

blizzard A storm with high wind, low temperature, and blowing snow. Strictly, in a blizzard the temperature is below 20°F (-6.7°C), wind speed is at least 35 mph (56 km/h), and falling or blowing snow reduces visibility to less than 440 yards (400 m).

buran A fierce, cold, northeasterly wind that blows across Russia and central Asia.

Calvin cycle The "dark," or light-independent, stage of photosynthesis in which carbon from carbon dioxide is combined with hydrogen and oxygen to form sugar in a series of chemical reactions. The steps in the cycle were discovered by the American biochemist Melvin Calvin.

carnivore An animal that feeds exclusively on other animals.

chlorophyll The green pigment, found in most plants, that absorbs light energy. This is then used to drive the reactions of photosynthesis.

chloroplast A structure in the cells of green plants. It contains chlorophyll and is the site of photosynthesis. Chloroplasts have some DNA and produce some of their own proteins.

competitive exclusion principle The principle that two or more species cannot exist for long in the same environment if each of them uses the same resources in the same way, because one will prove more successful than the others and will eliminate them. The principle was first stated by the Russian ecologist G.F. Gause and is sometimes called Gause's principle.

consumer An organism that is unable to manufacture its own food from simple ingredients but must obtain it by eating (consuming) other organisms.

convection Transfer of heat through a liquid or gas by the movement of the liquid or gas.

coterie A social group of mammals, such as prairie dogs.

cutaneous respiration Breathing by absorbing oxygen through the skin. Amphibians absorb a substantial proportion of their oxygen in this way.

deciduous Seasonally shed, like the leaves of certain trees and the antlers of deer. The word is sometimes applied to structures, such as the scales of some fish, that are shed readily (although not seasonally).

dicot A flowering plant (angiosperm) in which the embryo has two, or occasionally more, seed leaves (cotyledons).

Dust Bowl A region of the Great Plains in the United States where severe drought in the 1930s led to vast areas of soil being blown away by the wind.

ecology The study of the relationships among living organisms in a defined area and between the organisms and the nonliving features of their surroundings.

ecosystem A community of living organisms and their nonliving environment within a defined area. This may be of any size. A forest may be studied as an ecosystem and so may a drop of water.

eluviation The removal of soluble substances from the upper layers of soil and their deposition at a lower level.

endotherm An animal in which a fairly constant internal body temperature is maintained by physiological means, such as sweating, shivering, and the contraction and dilation of blood vessels. Birds and mammals are endotherms.

equinox One of the two occasions each year when the Sun crosses the equator and day and night are of equal length.

fungus A soft-bodied organism that obtains nutrients by absorbing them from its surroundings. Fungi are neither plants nor animals but constitute a kingdom of their own, the Fungi.

gill 1 The organ with which an aquatic animal obtains oxygen from water. It consists of thin membranes with a large surface area over which water flows. Oxygen passes from the water through the walls of blood vessels in the gill membrane and into the blood. Most aquatic animals have two gills. **2** A bladelike structure in the fruiting body of a fungus (often the visible stage in the life of the fungus, such as a toadstool or mushroom) from which spores are released.

ground water Water below ground that fills all the spaces between soil particles, thus saturating the soil.

gymnosperm A plant in which the ovule, bearing seeds, is not enclosed in an ovary but is carried naked on the surface of a modified leaf. In most gymnosperms the modified leaves form the scales of cones.

herbivore An animal that feeds exclusively on plants.

illuviation The deposition of substances in a layer of soil, usually in a lower layer.

insectivore An animal that feeds mainly or exclusively on insects.

invertebrate An animal that does not have a backbone.

lichen A plantlike organism consisting of a fungus and either an alga or a cyanobacterium (a bacterium that carries out photosynthesis) living in close association. The visible part of a lichen may be crustlike, scaly, leafy, or shrubby.

ligule A membrane covering a leaf, or a scaly outgrowth from a leaf.

line squall Storms, accompanied by rapid changes in wind direction and speed and often with lightning and thunder, that occur in a line, usually along an advancing cold front where cold air is undercutting and lifting warm, moist air.

lung The organ of respiration in air-breathing vertebrates. In land-dwelling mollusks (snails and slugs), the part of the body involved in respiration.

marsupial A mammal whose young are born at an early stage of their development and in many (but not all) species complete their development inside a maternal pouch (the marsupium). Kangaroos, wallabies, opossums, and koalas are marsupials.

monocot A flowering plant (angiosperm) in which the embryo has only one seed leaf (cotyledon). Grasses, palms, and plants that grow from bulbs are monocots.

mucus A sticky or slimy substance produced by some invertebrate animals and by special glands in vertebrates.

omnivore An animal that eats food derived from both plants and animals.

pampas Temperate grasslands in Argentina.

parasite An organism that lives on the surface, or inside the body, of another organism. The parasite is usually smaller than its host and gets food, shelter, or some other necessity from it. The effect of the parasite on its host may range from none at all to severe illness or even death.

paedogenesis Reproduction by an animal that is still in its larval stage.

pedology The scientific study of soils.

plane of the ecliptic An imaginary disk, the circumference of which is defined by the path traveled by the Earth in its orbit about the Sun.

poikilotherm An animal whose internal body temperature fluctuates according to the temperature of its surroundings. A fish is a poikilotherm.

prairie Temperate grassland in North America.

predator An organism that obtains food by consuming another organism. Most predators are animals that chase, overpower, and kill their prey, but insectivorous plants are also predators.

producer An organism, such as a green plant, that assembles large, complex substances from simple ingredients. These may then be eaten by consumers. On land the principal producers are green plants and in water they are single-celled plants called phytoplankton.

respiration 1 The oxidation of carbon to carbon dioxide in cells with the release of energy. **2** The action of breathing.

rhizome A plant stem that runs horizontally below the surface.

scavenger An animal that feeds on dead material, such as fallen plant or animal remains.

soil horizon An approximately horizontal soil layer that can be distinguished from the layers above and below it.

solstice The two dates each year when the Sun is directly overhead at one of the tropics at noon. This produces the longest hours of daylight in the hemisphere nearest the Sun (midsummer day) and the longest hours of darkness in the hemisphere farthest from the Sun (midwinter day). At present the solstices occur on about June 21 and December 22.

steppe Temperate grassland in eastern Europe, Russia, and central Asia.

stolon A plant stem that grows horizontally along the surface, sometimes called a "runner."

tiller A grass shoot that arises at ground level beside the other shoot(s).

tornado A narrow column of violent, spiraling wind, shaped like a funnel, produced by a large storm cloud.

transpiration The loss of water vapor through pores, called stomata in the leaves or lenticels in the stems, of green plants.

tropics Those parts of the world that lie between latitudes 23°30'N and 23°30'S. These latitudes mark the boundaries of the region within which the Sun is directly overhead at noon on at least one day each year. The Tropic of Cancer is to the north of the equator and the Tropic of Capricorn to the south.

veld Temperate grassland in South Africa.

vertebrate An animal that has a backbone. Vertebrates also have a bony skull containing the brain and a skeleton made from bone or cartilage. Fish, amphibians, reptiles, birds, and mammals are vertebrates.

water table The uppermost margin of the ground water, below which the soil is saturated and above which it is not, although it is wet.

weathering The physical and chemical processes by which rocks and minerals are broken down to particles of varying sizes, and soluble compounds are released into water.

Further Reading

Basics of Environmental Science by Michael Allaby. Routledge.

Biology by Neil A. Campbell. The Benjamin/Cummings Publishing Co. Inc.

The Encyclopedia of Birds edited by Christopher M. Perrins and Alex L.A. Middleton. Facts on File.

The Encyclopedia of Insects edited by Christopher O'Toole. Facts on File.

The Encyclopedia of Mammals edited by David Macdonald. Facts on File.

The Encyclopedia of Reptiles and Amphibians edited by Tim Halliday and Kraig Adler. Facts on File.

Flowering Plants of the World edited by V.H. Heywood. Oxford University Press, New York.

Green Planet edited by David M. Moore. Cambridge University Press.

The Hunters by Philip Whitfield. Simon and Schuster.

Hutchinson Encyclopedia of the Earth edited by Peter J. Smith. Hutchinson.

The Lie of the Land edited by K.J. Gregory. Oxford University Press, New York.

Longman Illustrated Animal Encyclopedia edited by Philip Whitfield. Guild Publishing.

The Oxford Encyclopedia of Trees of the World edited by Bayard Hora. Oxford University Press, New York.

Planet Earth: Cosmology, Geology, and the Evolution of Life and Environment by Cesare Emiliani. Cambridge University Press.

Snakes of the World by Chris Mattison. Blandford Press Ltd.

The Science of Ecology by Richard Brewer. Saunders College Publishing, Harcourt Brace College Publishers.

Web site:

Grand Prairie Friends, with links to other sites, is at: http://www.prairienet.org/gpf/homepage.html

Photographic Acknowledgments

6–7 Robert Harding Picture Library; **9** W. Carlson/Frank Lane Picture Agency; **15** Ron Petocz/World Wildlife Fund Photo Library; **17** Andromeda Oxford Limited; **21** John Shaw/Natural History Photographic Agency; **22** Tim Fitzharris; **24–25** Jeff Foott/Survival Anglia/Oxford Scientific Films; **27** W. Ervin/Natural Imagery; **32–33** Jeff Foott/Survival Anglia/Oxford Scientific Films; **41** Frank Lane Picture Agency; **42** John Shaw/Natural History Photographic Agency; **49** The Stock Market; **51** Jurgensfot; **52–53** The Stock Market; **54–55** Nicholas DeVore III/Bruce Coleman Limited; **Cover pictures:** *top*: Fritz Prenzel/Bruce Coleman Limited; *bottom*: David Hughes/Bruce Coleman Limited; *globe motif*: Terra Forma™ Copyright© 1995–1997 Andromeda Interactive Ltd.

While every effort has been made to trace the copyright holders of illustrations reproduced in this book, the publishers will be pleased to rectify any omissions or inaccuracies.

Set Index

Page numbers in *italics* refer to illustrations; volume numbers are in **bold**.